Fault Diagnosis
of Digital Circuits

Fault Diagnosis
of Digital Circuits

V. N. YARMOLIK

Minsk Radio Engineering Institute, USSR

JOHN WILEY & SONS

Chichester · New York · Brisbane · Toronto · Singapore

Other Wiley Editorial Offices

John Wiley & Sons, Inc., 605 Third Avenue,
New York, NY 10158-0012, USA

Jacaranda Wiley Ltd, G.P.O. Box 859, Brisbane,
Queensland 4001, Australia

John Wiley & Sons (Canada) Ltd, 22 Worcester Road,
Rexdale, Ontario M9W ILI, Canada

John Wiley & Sons (SEA) Pte Ltd, 37 Jalan Pemimpin   05-04,
Block B, Union Industrial Building, Singapore 2057

Library of Congress Cataloging-in-Publication Data:

IArmolik, V. N. (Vīacheslaw Nikolaevich)
 [Kontrol ' diagnostika tsifrovykh uzlov ĖVM. English]
 Fault diagnosis of digital circuits / V. N. Yarmolik.
  p. cm.
 Translation of: Kontrol 'i diagnostika t̄sifrovykh uzlov ĖVM.
 Includes bibliographical references.
 ISBN 0 471 92680 9
 1. Digital integrated circuits—Testing.   I. Title.
TK7874.I2413   1990
621.381'5—dc20   89-49651

British Library Cataloguing in Publication Data:

IArmolik, V. N. (Viacheslaw Nikolaevich)
Fault diagnosis of digital circuits.
1. Digital circuits
I. Title
621.38153

ISBN   0 471 92680 9

Typeset by Thomson Press (India) Ltd, New Delhi
Printed in Great Britain by Biddles Ltd, Guildford

CONTENTS

PREFACE

A distinctive feature of the present stage in computer equipment development is the continuous increase in functionality and complexity of computer components. Production procedures and techniques for electronic equipment are constantly being improved and their capabilities expanded. The use of advanced technologies increases the integration of digital components in computers, and extends their range.

The ever-growing complexity of digital components used in computers places more stringent requirements upon their reliability. Reliability may be improved by a complex mix of technological, maintenance and organizational measures. Among their diversity, technical diagnostic techniques for digital computer components are exceptional. One method of diagnosing computer parts and components is on-line testing. Its application at the stage of component manufacturing allows one to determine whether their behaviour is correct and to perform the fault location procedure, which eventually improves the basic reliability parameters (availability, maintainability and serviceability).

The principal emphasis of the present study has been on the use of up-to-date on-line testing techniques for digital devices employed in computers, which are most popular but have not been adequately covered in the literature.

Chapters 1–3 deal with the classical problems of on-line testing, pointing out the limited application of conventional approaches to the problem of diagnosing digital devices that employ large-scale and very-large-scale integrated (LSI and VLSI) chips. Consideration has been given to the theory of testable design for digital circuits, with this lead holding much promise for technical diagnostics. The concepts of testability and testability criteria have been introduced, and approaches to reduce the test generation time and test application complexity have been covered.

Chapters 4–7 cover compact testing methods, which are extensively used for diagnosing complex digital circuits, including LSI/VLSI circuits. Here classification of compact testing methods and their basic properties has been given. Among the variety of present-day compact testing methods, the most prominent are those for LSI/VLSI testing and self-testing. Random and pseudorandom testing are treated from the practical point of view. It has been noted that a more promising approach than pseudorandom testing for LSI self-testing is exhaustive testing, which allows one to solve the on-line testing methods most efficiently in conjunction with design for testability methods. The design procedures for a pseudorandom test sequence generator are discussed for both random and exhaustive testing.

Chapters 8 and 9 analyse the techniques of compressing the output responses of a digital circuit. It has been shown here that the compression technique most used is signature analysis. Specific features of signature analyser implementations that ensure the required diagnosis efficiency have been investigated. Methods of constructing parallel and multi-functional signature analysers as well as approaches to their use in fault localization have been proposed.

Chapter 10 surveys the most promising among recent signature generation techniques for binary sequences. Among their variety, the spectral method of estimating output responses of digital computer units seems to hold the lead.

Chapter 11 covers multi-output digital circuits. Consideration has been given to specific features of multi-line analysers. Parallel signature analyser design and application for digital testing have been discussed.

The author is sincerely grateful to E. S. Sogomonyan, Doctor of Technical Science, and to the reviewers, V. A. Sklyarov, Doctor of Technical Science, and J. A. Novikov, Candidate of Science, for their valuable contributions towards the enhancement of this book. He is also indebted to his colleagues from the Computer Department of the Minsk Radio Engineering Institute for their assistance in the preparation of the manuscript.

1

DIAGNOSTIC TESTING PROBLEMS

1.1. TYPES OF DIGITAL CIRCUIT FAULTS

The manufacture and maintenance of digital circuits invariably give rise to the problem of testing and diagnosing faults in them. The problem involves some interrelated tasks. The first is to determine the state of the circuit under test. It is common practice to distinguish between several types of digital circuit behaviour. The basic state of a digital circuit is good, i.e. the circuit behaviour meets the documented specifications. Otherwise, the circuit is considered to be in any of the various faulty conditions. When a circuit is found to be faulty, the second problem arises, i.e. fault (defect) localization, whose objective is to establish fault location and type.

Faults in a digital circuit may occur due to defective components such as logic gates that implement simple logic functions, memory elements, etc. Also a fault may occur as a result of breaks or shorts in interconnections, abnormal service conditions, design-rule and manufacturing violations, and for a number of other reasons [1–3].

Among the various faults there is a class of logic faults that cause the logic functions of a digital circuit element to change. This class of faults dominates other digital circuit faults. The mathematical models used for their description are the stuck-type fault, short-circuit fault, inverse fault and mix-up fault.

The most common model used for logic faults is the stuck-at fault: stuck-at-0 (or $\equiv 0$) and stuck-at-1 (or $\equiv 1$). It assumes that either a logic 0 or a logic 1 is fixed at one of the inputs or the output in a faulty logic element (gate). Such a fault model is referred to as classic and is widely used for defining other fault types.

A short-circuit (bridging) fault occurs when the inputs and outputs of a gate are shorted. Depending on the type of gates used, the faults have different effects on a digital circuit. Thus, a bridging fault between nodes x_1 and x_2 in a three-input AND gate (Fig. 1.1(a)) is equivalent to inserting a dummy two-inputs AND gate that interconnects the specified nodes. A similar bridging fault effect is valid for gates that use positive logic, and when an AND gate fails the logic function implemented by the gate remains unchanged. However, a similar fault in a three-input OR gate (Fig. 1.1(a)) changes the function implemented by it (Fig. 1.1(b)), which takes the form $f = x_1 x_2 + x_3$. When negative logic is used, the effect of a bridging fault between inputs of a gate is the insertion of a dummy two-input OR gate, thereby changing the function implemented by the AND gate.

Figures 1.1(b) and (c) show the effect of the above faults between nodes x_1 and x_2 in AND and OR gates (Fig. 1.1(a)) with positive and negative logic, respectively.

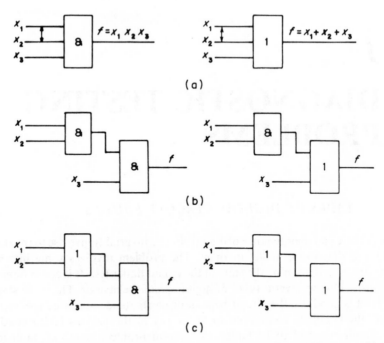

Fig. 1.1. (a) Examples of a bridging fault between lines x_1 and x_2 of AND and OR gates and its effect for (b) positive and (c) negative logic

Bridging faults may be classified into two types: input bridging and feedback bridging. For a combinational circuit implementing $F(x_1, x_2, \ldots, x_n)$, bridging between μ input lines of the circuit causes a μ-multiply short-circuit fault to occur. Figure 1.2(a) shows the logic model of a short-circuit fault between μ inputs of the circuit using positive logic. Figure 1.2(b) shows the logic model of a fault caused by bridging between the output line of the circuit and its μ inputs. The presence of such a connection between the output line and the input lines can cause a combinational circuit to oscillate or convert it to a sequential circuit. Thus for the circuit of Fig.

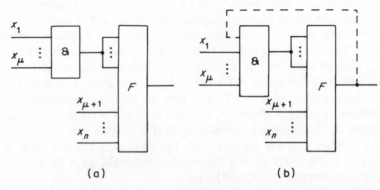

Fig. 1.2. Models of faults caused by a short-circuit (a) between μ inputs or (b) between the circuit output and its μ inputs

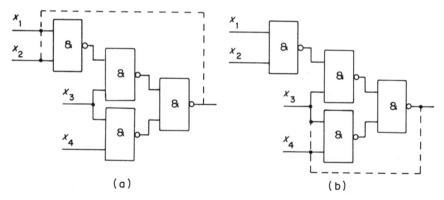

Fig. 1.3. Examples of a short-circuit fault (a) between output line and inputs x_1 and x_2 for $x_1 = x_2 = x_3 = 1$ and $x_4 = 0$ and (b) between output line and inputs x_3 and x_4 for $x_3 = x_4 = 1$

1.3(a), a short-circuit between the output line and inputs x_1 and x_2 for $x_1 = x_2 = x_3 = 1$ and $x_4 = 0$ causes the circuit to generate a high-frequency oscillation, whereas a short-circuit between the output and inputs x_3 and x_4 for $x_3 = x_4 = 1$ converts the circuit under consideration to a generator of an asynchronous sequence.

In the general case, a combinational circuit implementing $F(x_1, x_2, \ldots, x_n)$ that has a short-circuit fault of multiplicity μ between its output line and μ input lines of the circuit is converted to a sequential circuit if the set of variables $x_1, x_2, \ldots, x_n$ satisfies the condition

$$x_1 x_2 \ldots x_\mu \bar{F}(0, 0, \ldots, 0, x_{\mu+1}, \ldots, x_n) F(1, 1, \ldots, 1, x_{\mu+1}, \ldots, x_n) = 1. \tag{1.1}$$

Subject to the condition that

$$x_1 x_2 \ldots x_\mu F(0, 0, \ldots, 0, x_{\mu+1}, \ldots, x_n) \bar{F}(1, 1, \ldots, 1, x_{\mu+1}, \ldots, x_n) = 1. \tag{1.2}$$

a combinational circuit is converted to a high-frequency oscillator network [1].

For an example of a circuit with the discussed fault (Fig. 1.3(a)), by using equation (1.2) and the analytic description of its function $F = x_3(\bar{x}_1 + \bar{x}_2 + x_4)$ we obtain that $x_1 = x_2 = x_3 = \bar{x}_4 = 1$ is the condition for a circuit to oscillate. On the other hand, $F(x_1, x_2, 0, 0) = 0$, $F(x_1, x_2, 1, 1) = 1$ and hence $x_3 = x_4 = 1$ are the conditions for converting the discussed circuit to a sequential one (Fig. 1.3(b)).

Studies have proved that any bridging fault in a non-redundant digital circuit* can be detected. The conditions that allow one to generate a universal test for detecting all stuck-at faults at input and output lines of a digital circuit as well as all possible bridging between its outputs and inputs are presented in [4]. The complexity of detecting all possible short-circuit faults is explained by their large number. Thus, for a digital circuit that has n lines, the number of short-circuit faults of multiplicity μ will be represented by a value C_n^μ, which is rather large for practical values of n and μ [4].

An inverse fault is associated with a physical circuit defect that causes a dummy

---

*A combinational circuit is said to be non-redundant if, when good, it is defined by a Boolean function that is not the same as the one for its faulty condition [1].

inverter to appear at an input or output of a gate that makes up part of the circuit. In a number of cases inverse faults together with stuck-at faults are used [5] to construct a complete model of a faulty digital circuit. In this case all possible faults can be represented only by the models for stuck-at and inverse faults.

A mix-up fault occurs due to false connections in a digital circuit and is caused by design-rule violations or manufacturing defects that affect the functions performed by the circuit. As a rule, the number of faults of this kind is quite large and their detection amounts to finding stuck-at, bridging and inverse faults.

There exist a number of fault models that relate to specific technologies for digital circuits. Some of the models and the efficiency of their use are discussed in [1, 6–8].

Specific physical defects of very-large-scale integrated (VLSI) circuits and their mathematical models are discussed in [9–12], with the results presented in [11] and [12] relating to a single VLSI class, namely programmable logic arrays.

1.2. CONTROLLABILITY, OBSERVABILITY AND TESTABILITY OF DIGITAL CIRCUITS

The method of analysing digital circuits for testability is a major problem to be solved in their design and manufacture. With increase of integration, the cost of testing and diagnosing digital circuits has grown steadily. In a number of cases it amounts to 40–50% of their total cost [13]. It has been shown that the procedures of test generation and fault simulation are becoming more time- and labour-consuming. Thus, the density versus test generation time diagram (Fig. 1.4) [1] shows that, for VLSI technology, it will take many years for us to solve the problem. To minimize the complexity of digital circuit testing, a testability measure must be introduced by which the complexity of test generation and diagnostic procedures in circuits can be quantified at the design stage.

Engineering experience and confidence suggest that a testable circuit has the following properties [1]:

(i) The circuit can be easily put in the desired initial state.

(ii) Any internal state of the circuit can easily be set by the application of a test pattern to the primary inputs of the circuit.

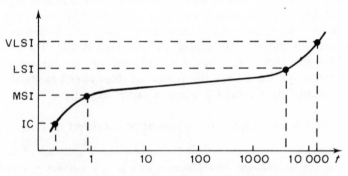

Fig. 1.4. Density versus test generation time

(iii) The internal state of the circuit can be uniquely identified through the primary outputs of the circuit.

The first two requirements explain the concept of circuit controllability, which refers to the ability to apply test patterns to a subcircuit via the primary inputs of the circuit. The third requirement is associated with circuit observability, which in conjunction with controllability directly affects the value used as a testability measure.

Among the many approaches for measuring controllability, observability and testability, we may distinguish the two major ones: deterministic and probabilistic [14]. As an example of the deterministic method, consider the theory of one testability analysis system, known as CAMELOT (Computer-Aided MEasure for LOgic Testability) [15].

In CAMELOT, controllability values Q are constrained to be in the range 0 to 1 for any circuit node. The maximum controllability value $Q = 1$ represents a primary input of the circuit, where it is as easy to establish a logic 0 as it is a logic 1. Controllability value $Q = 0$ represents any circuit node which cannot be set to 0 or 1. In practice, the majority of nodes in a circuit will have $0 < Q < 1$. Generally, the controllability value for an n-input combinational circuit can be evaluated [15] as

$$Q_0 = Q_t \frac{1}{n} \sum_{i=1}^{n} Q_i \tag{1.3}$$

where Q_0 and Q_i are controllability values for the output and inputs nodes in a circuit, respectively. Q_t is a measure of the complexity in generating an output value of 0 or 1 for $Q_i = 1$, $i = 1$ to n. By the technique presented in [15],

$$Q_t = 1 - \left| \frac{F(0) - F(1)}{F(0) + F(1)} \right|. \tag{1.4}$$

Here $F(0)$ and $F(1)$ are, respectively, the total number of 0 and 1 values of function $F(x_1, x_2, \ldots, x_n)$ implemented by a combinational circuit. For simple digital circuits, the values of $F(0)$ and $F(1)$ can be obtained from the truth tables for the circuit. However, $F(0) = 2^n - F(1)$ is improbable for all cases since the function sometimes generated by the circuit is not completely defined and some input combinations of variables are forbidden.

Figure 1.5 shows examples of controllability calculations for two-input logic gates NOT, AND, OR and EOR. The procedure for calculating the controllability for synchronous circuits with memory elements as well as circuits with feedback loops is slightly more complex. A simple example of a circuit with a feedback loop is an

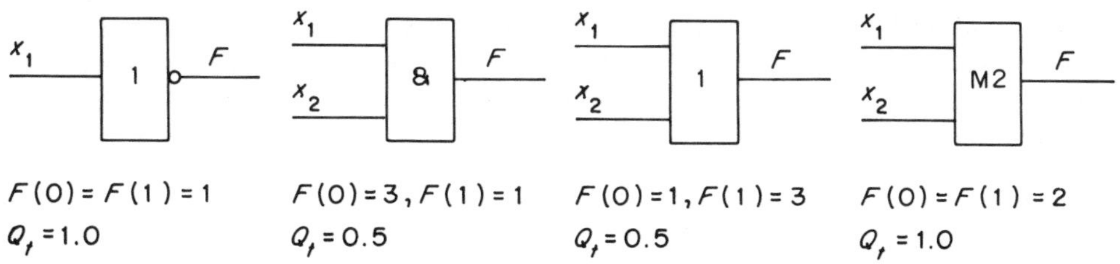

$F(0) = F(1) = 1$ $F(0) = 3, F(1) = 1$ $F(0) = 1, F(1) = 3$ $F(0) = F(1) = 2$
$Q_t = 1.0$ $Q_t = 0.5$ $Q_t = 0.5$ $Q_t = 1.0$

Fig. 1.5. Calculation of controllability for two-input logic gates NOT, AND OR and EOR

asynchronous R–S flip-flop for which $Q_1 = Q_2 = 1$. By using equation (1.3) and assuming that Q_{01} and Q_{02} are unknown values, we obtain the system of equations

$$Q_{01} = 0.5(1 + Q_{02})/2$$

$$Q_{02} = 0.5(1 + Q_{01})/2$$

where 0.5 denotes Q_t of a two-input NAND gate as per equation (1.4). The solution of the system is $Q_{01} = Q_{02} = 0.33$.

Observability O for a circuit, in the same way as controllability, is constrained to be in the range 0 to 1 in CAMELOT. The condition $O = 0$ means that there is no way of propagating fault–effect data from its origin to a circuit output. Also, the condition $O = 1$ denotes that fault–effect data are always present in a circuit output irrespective of its input variables. The same as controllability, observability for the majority of circuit nodes lies in the range $0 < O < 1$ and can be calculated by an analytic relationship

$$O_0 = O_i O_t Q_{tn} \tag{1.5}$$

where O_i is fault observability in its origin i, O_0 is fault observability on a circuit output, O_t is a measure of the ease with which the fault propagates from its origin to a circuit output and Q_{tn} reflects the integrated estimation of controllability of circuit inputs that ensures the propagation of a fault to an output (path sensitization condition). The Q_t value is calculated by

$$O_t = P_1/(P_1 + P_2) \tag{1.6}$$

where P_1 and P_2 are the total number of sensitive and insensitive paths, respectively, in the circuit under test. The value O_t for a NOT gate is 1 since $P_1 = 1$ and $P_2 = 0$. At the same time, O_{t1}, i.e. observability of a fault propagated from the first input of a two-input AND gate to its output, is 0.5 since $P_1 = P_2 = 1$. For a three-input AND gate, the observability value is 0.25 since $P_1 = 1$ and $P_2 = 3$; therefore $O_{t1} = O_{t2} = O_{t3} = 0.25$ by symmetry.

The value Q_{tn} reflecting controllability of inputs in an asynchronous combinational circuit that ensures sensitivity of a selected path can be calculated by

$$Q_{tn} = \frac{1}{n} \sum_{i=1}^{n} Q_i \tag{1.7}$$

where Q_i is controllability of the ith input node and n is the total number of nodes.

By using the concepts of observability and controllability for asynchronous combinational circuits discussed above, the CAMELOT system allows one to estimate the said properties for a digital circuit with memory elements, fan-in and fan-out branches as well as feedback loops. From controllability Q and observability O we can obtain a value of testability $T_k = f(Q, O)$. For a particular node, testability can be estimated as $T_k = QO$, and the overall testability value for the circuit is estimated by [14]

$$T_c = \frac{\sum_{k=1}^{K} T_k}{K}$$

where T_k is testability for the kth node in a circuit and K is the total number of nodes.

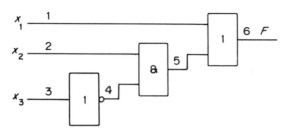

Fig. 1.6. Combinational circuit used in an example of a testability calculation

Consider an example of calculating the testability value for a simple combinational circuit (Fig. 1.6). First, by using equations (1.3)–(1.6) we obtain the controllability and observability values for every node in the circuit under test (Table 1.1). The final value that reflects the circuit testability is $T_c = 0.35$, quantifying the measure of complexity in fault detection for the circuit.

Unlike the majority of deterministic methods of calculating controllability, observability and testability for a combinational circuit, probability methods allow one to estimate these properties for a particular node in a circuit for both logic 0 and 1 [16]. Thus controllability Q_{k1} of a circuit node for a stuck-at-1 fault ($f_k \equiv 1$) is calculated as the ratio of input vectors for which logic variable is 0 on the kth node to the total number of input vectors. Similarly, we can find the Q_{k0} value for $f_k = 0$. Observability of a stuck-at-1 fault Q_{k1} (stuck-at-0 fault Q_{k0}) for a node k is calculated as the ratio of input vectors for which the kth node fault $\equiv 1(\equiv 0)$ propagates to the circuit input to the total number of vectors.

Considering the above definitions, we can represent a number of properties for these responses. It is evident that

$$0 \leqslant Q_{k1}, Q_{k0}, O_{k1}, O_{k0} \leqslant 0$$

besides $Q_{k1} = 1 - Q_{k0}$ and $O_{k0} = O_{k1}$.

As an example of defining these responses, consider node 5 (Fig. 1.6) with a fault $f_5 \equiv 0$. The value of function $f_5 = 1$ is ensured by two input patterns $x_1 x_2 x_3$, 010 and 110, from a set of input patterns whose cardinality is 8. Therefore, $Q_{50} = 0.25$ and the value of O_{50} is calculated as the ratio 4/8, i.e. $O_{50} = 0.5$. The present case is based on the assumption that the condition of path sensitization from node 5 to node 6 is ensured by an input pattern $x_1 x_2 x_3 = 0XX$, where X may be 0 or 1.

In view of the fact that controllability and observability are the only arguments

Table 1.1.

k	Q_k	O_k	T_k	T_c
1	1	0.75	0.75	
2	1	0.25	0.25	
3	1	0.25	0.25	≈ 0.35
4	1	0.25	0.25	
5	0.5	0.5	0.25	
6	0.375	1	0.375	

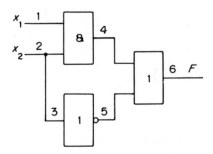

Fig. 1.7. A combinational circuit

of a function that defines testability for a circuit, we can calculate testability T_{50} as a product of arguments, i.e. $T_{50} = Q_{50}O_{50} = 0.25 \times 0.5 = 0.125$. A comparison between the obtained result and testability for the circuit node in Table 1.1 reveals that testability may vary with the approach used. Because of this, different functional dependences between testability and controllability/observability have been used to obtain a more adquate testability estimate. Also, geometric dependences [17, 18], root-mean-square estimates and a product with weighting coefficients [15] were used for the purpose.

A number of examples show that testability has practically no dependence on controllability and observability for some circuits. Thus for the circuit of Fig. 1.7 and its fault $\equiv 1$ on the second input of a two-input AND gate, controllability Q_{21} is 0.5 and observability O_{21} is 0.25. It is evident that the values obtained for these properties are high enough to ensure a reasonable testability value of T_{21}. However, as the analysis of the circuit under test shows, there is no input pattern $x_1 x_2$ that allows one to detect a fault on the second node of the two-input AND gate. This clearly demonstrates that $T_{21} = 0$ [16]. The contradiction is explained by the fact that the circuit of Fig. 1.7 is redundant because the following evident identity, $x_1 x_2 + \bar{x}_2 = x_1 + \bar{x}_2$, is satisfied.

The second example that demonstrates complex functional dependence between testability and observability/controllability is the circuit of Fig. 1.8. The circuit properties are $Q_{50} = 0.25$ and $O_{50} = 0.75$. In this case, there is a fault $\equiv 0$ on node 5 of the circuit for which there exists only one input test pattern $x_1 x_2 = 11$ of four test patterns that may be used for fault detection. This implies that T_{50} is significantly higher than the product $Q_{50}O_{50}$.

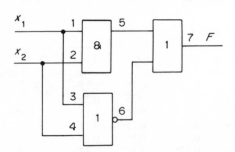

Fig. 1.8. A redundant combinational circuit

The above examples, the second one in particular, allow one to define testability more rigorously. Testability $T_{k1}(T_{k0})$ for a fault $\equiv 1(\equiv 0)$ to appear at the kth node of a combinational circuit is calculated as the ratio of test patterns that guarantee fault detection to the total number of test patterns.

Because it is very difficult and even impracticable to obtain the values of T_{k1} and T_{k0} in accordance with the above definition, we can estimate testability in terms of controllability and observability only. This complies with inequalities

$$T_{k1} \leqslant \min(Q_{k1}, O_{k1}) \qquad T_{k0} \leqslant \min(Q_{k0}, O_{k0})$$

that are valid for the general case. However, a rigorous relationship of T_{k1} and T_{k0} to $Q_{k1}O_{k1}$ and $Q_{k0}O_{k0}$ can be obtained only for the special case of a tree circuit, for which the following equalities hold:

$$T_{k1} = Q_{k1}O_{k1} \qquad T_{k0} = Q_{k0}O_{k0}.$$

Irrespective of the computational complexity of testability and the absence of established procedures for measuring it, in practice there are many testability analysis systems for digital circuits that eventually allow one to solve the problem of diagnosing computer units efficiently. Among the systems that implement the concepts underlying the CAMELOT system discussed, the RMES system [18] estimates the integral testability value for the two probable faults $\equiv 0$ and $\equiv 1$ on the specified node of a digital circuit. Various testability measures are defined in the TEST/80 [14], SCOAP and COMET [19] systems that allow one to estimate testability for $\equiv 0$ and $\equiv 1$ independently. VICTOR [20] (VLSI, Identifier of Controllability, Testability Observability and Redundancy) allows one to estimate redundancy of the VLSI design and to give recommendations as to its layout modification.

2

TEST GENERATION

2.1. ONE-DIMENSIONAL PATH SENSITIZATION

In general, test generation must resolve two major problems: to ensure fault observability in its origin and to propagate the fault effect to one of the primary outputs in the circuit. At present, there are different approaches and methods available for implementing both heuristic and rigorous formal algorithms.

Historically, one of the first approaches used for test generation was one-dimensional path sensitization [21]. The principle of the approach is to sensitize a one-dimensional path from the origin of the failure to propagate its effect to an output of the circuit. The sensitization procedure is as follows:

(i) Derive a test for the specified fault to appear in its origin. Thus, for example, the observability condition for a stuck-at-1 fault is a logic 0 on the circuit line where the fault has occurred.

(ii) Choose a path over which a fault propagates to the circuit output. It is specified as a sequence of elements that belong to the path.

(iii) The chosen path sensitivity is determined in terms of its input variable elements. For a chosen path element, its input variables are specified so that its output value is determined by the input line connected to the preceding gate output in the path.

(iv) Calculate the values of input variables of the circuit that will determine fault observability and its propagation to an output of the circuit, i.e. the fault detection condition in terms of its input variables. The calculation result is a sought-for test.

Steps (i) to (iii) of the algorithm are often called the forward-trace phase and step (iv) is called the backward-trace phase.

As an example of a one-dimensional path sensitization technique, let us consider the digital circuit of Fig. 2.1. Let us derive a test for a $\equiv 1$ fault on the output of gate 4.

By the above technique, the specified fault will be observed with $f_4 = 0$ or, taking account of the function performed by gate 4,

$$f_4(x_4, f_2(x_5, x_6), x_7) = x_4(x_5 + x_6)x_7 = 0.$$

Then choose a path through gates 5 and 6. The path sensitization through gate 5 is determined by the value $f_1 = 0$ on its second input, and therefore f_5 changes with any change of f_4. With $f_3 = 1$, a path can be sensitized via gate 6. As a result of the forward trace of the one-dimensional path sensitization, we obtain the observability of a fault in its origin and its transfer to the circuit output in the form of the system of logic equations

$$f_4 = 0 \qquad f_1 = 0 \qquad f_3 = 1. \tag{2.1}$$

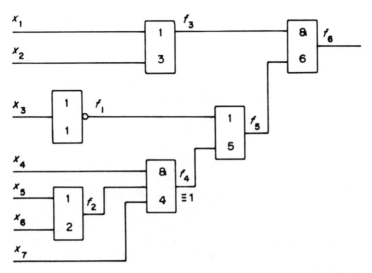

Fig. 2.1. A digital circuit illustrating a one-dimensional path sensitization

The solution to the system (2.1) is essentially a procedure performed at backward tracing of the one-dimensional path sensitization. With account of functions implemented by the elements of the circuit, the system takes the form

$$x_4 x_5 x_7 + x_4 x_6 x_7 = 0 \qquad \bar{x}_3 = 0 \qquad x_1 + x_2 = 1.$$

A multitude of solutions, each being a test for the specified fault, will obey the resulting system of equations. In fact, for a vector of variables $x_1 x_2 x_3 x_4 x_5 x_6 x_7 = 1110111$

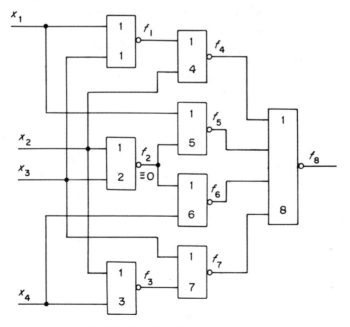

Fig. 2.2. The Schneider counterexample

that represents one of the possible solutions, the value f_6 on the output line of the circuit of Fig. 2.1 will be 1 for $f_4 \equiv 1$ or 0 for the fault-free circuit. This is indicative of fault observability on an output of the circuit.

However, irrespective of its evident simplicity, the one-dimensional path sensitization technique has not found wide application since the forward- and backward-trace phases discussed above do not necessarily produce the sought-for test. This is apparent from the classical Schneider counterexample (Fig. 2.2) [22]. In this case it is impossible to derive a test for $f_2 \equiv 0$ by using the one-dimensional path sensitization technique. For $f_2 \equiv 0$ to be observable in its origin, variables x_2 and x_3 must be equal to 0, and its propagation to an output of the circuit is ensured by $x_4 = 0$ and $f_4 = f_5 = f_7 = 0$. For $f_5 = 0$, x_1 must be equal to 1 and therefore $f_1 = 0$, subject to $x_2 = 0$, and we obtain $f_4 = 1$, which conflicts with the condition $f_4 = 0$ for a path to be sensitized through gate 8. Similarly, we can demonstrate that another sensitization along a path through gates 5 and 8 must also fail. At the same time, it is evident that the relation $x_1 = x_4 = 0$ allows one to sensitize two paths at a time for propagating $f_2 = 0$ and the input pattern $x_1 x_2 x_3 x_4 = 0000$ provides a test for the fault in question.

The above example shows that, in order to solve the problem of generating test patterns, we must sensitize many paths at a time. Let us consider the multiple-path sensitization approach called the *d*-algorithm.

2.2. *THE d-ALGORITHM*

The multiple-path sensitization approach (*d*-algorithm) formulated by Roth [23] was the first algorithmic method of deriving tests for non-redundant combinational circuits. In other words, the said algorithm guarantees finding a test for the specified fault if such a test exists. Roth's *d*-algorithm is based on cubical definition of Boolean functions, which is one of the possible forms of their expression. Among cubic representations of Boolean functions used in the algorithm we can isolate the so-called singular cubes, the set of which is a singular cover for a circuit implementing a

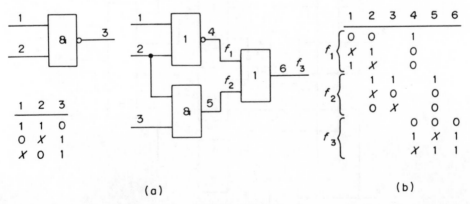

(a)   (b)

Fig. 2.3. A set of singular cubes for (a) the two-input NAND gate and (b) a circuit consisting of two-input AND, NOR and OR gates

Boolean function. For denoting singular cubes, symbols 0, 1 and X are used, where X means that a variable on the coordinate may assume either 0 or 1. Fig. 2.3(a) shows a set of singular cubes for a two-input NAND gate, and Fig. 2.3(b) shows a set of singular cubes for a more complex circuit consisting of a two-input AND, a two-input NOR and a two-input OR gate [1].

The singular cubes discussed are used to obtain d-cubes, which are represented by symbols 0, 1, X, d and $\bar{d}$. A d may assume either 0 or 1, and a $\bar{d}$ will have the opposite value. Thus all $\bar{d}$s assume the opposite value.

A d-cube representing the behaviour of a digital circuit can be constructed by intersecting two singular cubes from its singular cover that have different output values according to the following algebraic rules:

$$0 \cap 0 = 0 \qquad 0 \cap X = 0 \qquad X \cap 0 = 0$$
$$1 \cap 1 = 1 \qquad 1 \cap X = 1 \qquad X \cap 1 = 1$$
$$X \cap X = X \qquad 1 \cap 0 = d \qquad 0 \cap 1 = \bar{d}.$$

Thus, based on

$$C_1 = \frac{\begin{array}{ccc} 1 & 2 & 3 \end{array}}{\begin{array}{ccc} 1 & 1 & 0 \end{array}} \qquad \text{and} \qquad C_2 = \frac{\begin{array}{ccc} 1 & 2 & 3 \end{array}}{\begin{array}{ccc} X & 0 & 1 \end{array}}$$

from the singular cover of the two-input NAND gate we obtain the d-cube

$$C_1 \cap C_2 = \frac{\begin{array}{ccc} 1 & 2 & 3 \end{array}}{\begin{array}{ccc} 1 & d & \bar{d} \end{array}}$$

by coordinate intersection.

The primitive d-cubes of a fault are used to specify the existence of a given fault. The primitive d-cube consists of an input pattern that brings the effect of a fault to the output of the gate and symbol d or $\bar{d}$ on the output coordinate. For the circuit of Fig. 2.3(a) and a $\equiv 0$ fault on its output, the corresponding primitive d-cube is

$$\frac{\begin{array}{ccc} 1 & 2 & 3 \end{array}}{\begin{array}{ccc} 0 & 0 & d \end{array}}.$$

Here, the d is interpreted as a 1 on the output coordinate for the fault-free circuit and a 0 for the faulty circuit. At the same time, for cubes

$$\begin{array}{ccc} 1 & 2 & 3 \\ 1 & X & \bar{d} \\ X & 1 & \bar{d} \end{array}$$

that represent the two-input NOR gate behaviour with a $\equiv 1$ fault on its output, the $\bar{d}$ symbol denotes that its output should be 0 for the fault-free gate and 1 for a faulty gate.

It is a relatively simple matter to construct a d-cube of a fault for a single gate of the circuit with a stuck-at fault on its output. However, the d-algorithm objective is to construct a full circuit d-cube of the fault. For that purpose, we can use the d-intersection that has been introduced by Roth to formalize the multiple-path sensitization technique. As applied to the d-algorithm, the d-intersection of two d-cubes

$A = (a_1, a_2, \ldots, a_n)$ and $B = (b_1, b_2, \ldots, b_n)$, where $a_i, b_i \in \{0, 1, X, d, \bar{d}\}, i = 1$ to n, is given by

(i) $X \underset{d}{\bigcap} b_i = b_i; a_i \underset{d}{\bigcap} X = a_i$

(ii) if $a_i \neq X$ and $b_i \neq X$ then

$$a_i \underset{d}{\bigcap} b_i = \begin{cases} a_i & \text{if } b_i = a_i \\ \varnothing & \text{otherwise} \end{cases}$$

where $\varnothing$ is an empty set.

Finally, the d-intersection of cubes A and B is

$$A \underset{d}{\bigcap} B = \left(a_1 \underset{d}{\bigcap} b_1, a_2 \underset{d}{\bigcap} b_2, \ldots, a_n \underset{d}{\bigcap} b_n \right)$$

if for any i,

$$a_i \underset{d}{\bigcap} b_i \neq \varnothing.$$

Otherwise

$$A \underset{d}{\bigcap} B = \varnothing.$$

The proper d-algorithm consists of the following key stages.

(i) A primitive d-cube of the fault under test is chosen for a gate in the circuit.

(ii) All possible paths from the faulty gate to a primary output of the circuit are sensitized. This is done by intersecting the d-cube of the fault with d-cubes of gates that are connected successively between the original faulty gate and the circuit outputs. The primary intersection of the d-cube of the fault with the d-cube of a successor gate results in the d-cube of the fault under test, which represents the behaviour of a two-gate circuit. The process of intersecting d-cubes of the fault with d-cubes of the gates placed in the fault propagation path is called the d-drive [21, 24]. The d-drive lasts until an output coordinate has a d or $\bar{d}$.

(iii) The backtrack operation is performed as successive intersection of the resulting d-cube of the specified fault that represents the gate's fault behaviour and the d-cubes of the gates in the fault propagation path with singular cubes of other gates.

The application of the d-algorithm is demonstrated by detecting the fault $f_2 \equiv 0$ in the circuit of Fig. 2.4 [1].

First, we derive a set of singular cubes for gates 1 and 3, d-cubes for gates 4 and 5, and primitive d-cubes of the fault for gate 2:

f_1			f_2			f_3			f_4			f_5		
1	2	5	3	4	6	3	5	7	2	6	4	7	8	9
X	1	0	1	0	d	0	X	1	0	d	$\bar{d}$	1	$\bar{d}$	d
1	X	0	0	1	d	X	0	1	d	0	$\bar{d}$	$\bar{d}$	1	d
0	0	1	1	1	$\bar{d}$	1	1	0	d	d	$\bar{d}$	$\bar{d}$	$\bar{d}$	d

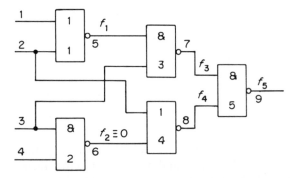

Fig. 2.4. Application of the *d*-algorithm for a circuit with an $f_2 \equiv 0$ fault

From the set of *d*-cubes of the fault for gate 2 we choose the cube

$$C_0 = \frac{\begin{array}{ccc} 3 & 4 & 6 \end{array}}{\begin{array}{ccc} 1 & 0 & d \end{array}}.$$

Then at the first step of the *d*-drive we perform *d*-intersection of the fault *d*-cube C_0 with the *d*-cube C_1 defining gate 4. As a result

	1	2	3	4	5	6	7	8	9
$C_0 =$	X	X	1	0	X	d	X	X	X

$$\underset{d}{\cap}$$

	1	2	3	4	5	6	7	8	9
$C_1 =$	X	0	X	X	X	d	X	$\bar{d}$	X
$C_2 =$	X	0	1	0	X	d	X	$\bar{d}$	X.

The *d*-cube C_2 intersects the *d*-cube C_3 of gate 5

	1	2	3	4	5	6	7	8	9
$C_2 =$	X	0	1	0	X	d	X	$\bar{d}$	X

$$\underset{d}{\cap}$$

	1	2	3	4	5	6	7	8	9
$C_3 =$	X	X	X	X	X	X	1	$\bar{d}$	d
$C_4 =$	X	0	1	0	X	d	1	$\bar{d}$	d.

The analysis of cube C_4 shows that a *d* present on node 9 of the circuit under test is indicative of the end of *d*-drive.

In the backtrack operation, cube C_4 is successively intersected with singular cubes of gates 3 and 1. As a result, we obtain

	1	2	3	4	5	6	7	8	9
$C_4 =$	X	0	1	0	X	d	1	$\bar{d}$	d

$$\underset{d}{\cap}$$

$$C_5 = \begin{array}{ccccccccc} X & X & X & X & 0 & X & 1 & X & X \end{array}$$

$$C_6 = \begin{array}{ccccccccc} X & 0 & 1 & 0 & 0 & d & 1 & \bar{d} & d \end{array}$$

$$\bigcap_{d}$$

$$C_7 = \begin{array}{ccccccccc} 1 & X & X & X & 0 & X & X & X & X \end{array}$$

$$C_8 = \begin{array}{ccccccccc} 1 & 0 & 1 & 0 & 0 & d & 1 & \bar{d} & d. \end{array}$$

Thus, the resulting cube C_8 demonstrates that with a pattern $x_1 x_2 x_3 x_4 = 1010$ applied to the inputs of the circuit and $f_2 \equiv 0$, if it exists, there are conditions for fault propagation to the circuit output. In other words, the input pattern 1010 is test for $f_2 \equiv 0$.

The d-algorithm just discussed allows one to generate a test for any fault if such a test exists. Besides, it can identify redundancy in a digital circuit [1], requires reasonable memory size and is practically rather efficient for wide use in automatic test generation systems.

A version of the d-algorithm that is currently widely employed in practice [25] is the LASAR (Logic Automated Stimulus and Response) system. Further practical development of the d-algorithm is presented in other publications [14, 26].

2.3. BOOLEAN DIFFERENCES

The present method is based on the use of Boolean differences (Boolean partial derivatives), which can be derived as follows. A Boolean partial derivative of a Boolean function $F(x_1,\ldots,x_n)$ with respect to variable x_i, $i \in \{1, 2, \ldots, n\}$ is called the Boolean function defined as

$$\frac{\partial F(x_1,\ldots,x_n)}{\partial x_i} = F(x_1,\ldots,x_i,\ldots,x_n) \oplus F(x_1,\ldots,\bar{x}_i,\ldots,x_n). \tag{2.2}$$

The following properties are valid for the function [1]:

(i) $\quad \dfrac{\overline{\partial F(x_1,\ldots,x_n)}}{\partial x_i} = \dfrac{\partial \overline{F(x_1,\ldots,x_n)}}{\partial x_i}$

where $\overline{F(x_1,\ldots,x_n)}$ denotes the negation of $F(x_1,\ldots,x_n)$ with x_i being its argument.

(ii) $\quad \dfrac{\partial F(x_1,\ldots,x_n)}{\partial \bar{x}_i} = \dfrac{\partial F(x_1,\ldots,x_n)}{\partial x_i}.$

(iii) $\quad \dfrac{\partial}{\partial x_i} \dfrac{\partial F(x_1,\ldots,x_n)}{\partial x_j} = \dfrac{\partial}{\partial x_j} \dfrac{\partial F(x_1,\ldots,x_n)}{\partial x_i}.$

(iv) $\quad \dfrac{\partial [F(x_1,\ldots,x_n) G(x_1,\ldots,x_n)]}{\partial x_i} = F(x_1,\ldots,x_n) \dfrac{\partial G(x_1,\ldots,x_n)}{\partial x_i}$

$$\oplus G(x_1,\ldots,x_n) \dfrac{\partial F(x_1,\ldots,x_n)}{\partial x_i} \oplus \dfrac{\partial F(x_1,\ldots,x_n)}{\partial x_i} \dfrac{\partial G(x_1,\ldots,x_n)}{\partial x_i}.$$

(v) $\dfrac{\partial[F(x_1,\ldots,x_n) + G(x_1,\ldots,x_n)]}{\partial x_i}$

$$= \overline{F(x_1,\ldots,x_n)}\dfrac{\partial G(x_1,\ldots,x_n)}{\partial x_i} \oplus \overline{G(x_1,\ldots,x_n)}\dfrac{\partial F(x_1,\ldots,x_n)}{\partial x_i}$$

$$\oplus \dfrac{\partial F(x_1,\ldots,x_n)}{\partial x_i}\dfrac{\partial G(x_1,\ldots,x_n)}{\partial x_i}.$$

(vi) $\dfrac{\partial F(x_1,\ldots,x_n)}{\partial x_i} = 0$

if $F(x_1,\ldots,x_n)$ is independent of x_i.

(vii) $\dfrac{\partial F(x_1,\ldots,x_n)}{\partial x_i} = 1$

if $F(x_1,\ldots,x_n) = x_i$.

(viii) $\dfrac{\partial[F(x_1,\ldots,x_n)G(x_1,\ldots,x_n)]}{\partial x_i} = F(x_1,\ldots,x_n)\dfrac{\partial G(x_1,\ldots,x_n)}{\partial x_i}$

if $F(x_1,\ldots,x_n)$ is independent of x_i, and hence for $F(x_1,\ldots,x_n) = x_1\cdots x_i\cdots x_n$

$$\dfrac{\partial F(x_1,\ldots,x_n)}{\partial x_i} = x_1\cdots x_{i-1}x_{i+1}\cdots x_n.$$

(ix) $\dfrac{\partial[F(x_1,\ldots,x_n) + G(x_1,\ldots,x_n)]}{\partial x_i} = \overline{F(x_1,\ldots,x_n)}\dfrac{\partial G(x_1,\ldots,x_n)}{\partial x_i}$

if $F(x_1,\ldots,x_n)$ is independent of x_i and respectively for $F(x_1,\ldots,x_n) = x_1 + \cdots + x_i + \cdots x_n$

$$\dfrac{\partial F(x_1,\ldots,x_n)}{\partial x_i} = \bar{x}_1\cdots\bar{x}_{i-1}\bar{x}_{i+1}\cdots\bar{x}_n.$$

Let us find a Boolean derivative of $F(x_1, x_2, x_3, x_4) = x_1x_2 + x_3x_4$ with respect to variable x_3 (Fig. 2.5). By property (v),

$$\dfrac{\partial(x_1x_2 + x_3x_4)}{\partial x_3} = \overline{x_1x_2}\dfrac{\partial(x_3x_4)}{\partial x_3} \oplus \overline{x_3x_4}\dfrac{\partial(x_1x_2)}{\partial x_3} \oplus \dfrac{\partial(x_1x_2)}{\partial x_3}\dfrac{\partial(x_3x_4)}{\partial x_3}.$$

Considering property (vi), the above equation may be rearranged to give

$$\dfrac{\partial(x_1x_2 + x_3x_4)}{\partial x_3} = \overline{x_1x_2}\dfrac{\partial(x_3x_4)}{\partial x_3}.$$

By sequentially applying properties (iv), (vi) and (vii), we finally obtain

$$\dfrac{\partial(x_1x_2 + x_3x_4)}{\partial x_3} = \bar{x}_1x_4 + \bar{x}_2x_4.$$

We can similarly define a partial derivative of $F(x_1,\ldots,x_n)$ with respect to variable f_k generated at an internal node of the circuit that realizes the function.

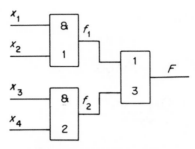

Fig. 2.5. A combinational circuit

An original function is represented as $F(x_1,\ldots,x_n, f_k)$ where f_k is also dependent on input variables $x_1,\ldots,x_n$. Thus, for the example of Fig. 2.5, function $F(x_1, x_2, x_3, x_4) = x_1 x_2 + x_3 x_4$ takes the form $F(x_1, x_2, f_2) = x_1 x_2 + f_2$ when a partial derivative is defined with respect to variable f_2 generated at the output of the second element. The derivative of the resulting function can be estimated by

$$\frac{\partial(x_1 x_2 + f_2)}{\partial f_2} = x_1 x_2 \frac{\overline{\partial f_2}}{\partial f_2} \oplus \overline{f_2} \frac{\partial(x_1 x_2)}{\partial f_2} \oplus \frac{\partial f_2}{\partial f_2} \frac{\partial(x_1 x_2)}{\partial f_2}.$$

By properties (vi) and (vii) we obtain

$$\frac{\partial(x_1 x_2 + f_2)}{\partial f_2} = \overline{x_1 x_2}.$$

By using the definition of the Boolean difference of $F(x_1,\ldots,x_n)$ with respect to variable x_i, we can show that subject to the equality

$$\frac{\partial F(x_1,\ldots,x_n)}{\partial x_i} = 1 \tag{2.3}$$

any change of x_i will cause a change in the value of function $F(x_1,\ldots,x_n)$; otherwise, the value of x_i has no effect on the function value. In other words, equation (2.3) is a condition for observability of a change in x_i on the output of the circuit that realizes $F(x_1,\ldots,x_n)$.

The discussed property of the Boolean difference is used to establish fault propagation conditions on generating a test pattern. Fault observability at its origin is determined by the equality of function f_k realized at this node to 1 for $\equiv 0$ or to 0 for $\equiv 1$. By combining both conditions we can construct a system of logical equations for $\equiv 0$ fault

$$f_k(x_1,\ldots,x_n) = 1$$

$$\frac{\partial F(x_1,\ldots,x_n, f_k)}{\partial f_k} = 1$$

and for $\equiv 1$ fault

$$f_k(x_1,\ldots,x_n) = 0$$

$$\frac{\partial F(x_1,\ldots,x_n, f_k)}{\partial f_k} = 1.$$

In practice, the discussed properties of the system of equations can be represented by a single equation

$$f_k(x_1,\ldots,x_n)\frac{\partial F(x_1,\ldots,x_n,f_k)}{\partial f_k} = 1 \qquad (2.4)$$

or

$$\overline{f_k(x_1,\ldots,x_n)}\frac{\partial F(x_1,\ldots,x_n,f_k)}{\partial f_k} = 1. \qquad (2.5)$$

The solution to these equations (2.4) and (2.5) is a test for detecting a fault $\equiv 0$ ($\equiv 1$) occurring at the kth node of the circuit under test that has been defined by function $F(x_1,\ldots,x_n,f_k)$. For the example of Fig. 2.5, the solution of equation (2.5) for $f_2 = x_3 x_4$ and $F = x_1 x_2 + f_2$ is a test for detecting the fault $f_2 \equiv 1$.

In view of $\partial F(x_1 x_2 + f_2)/\partial f_2 = x_1 x_2$, we obtain

$$\overline{x_3 x_4}\,\frac{\partial F(x_1 x_2 + f_2)}{\partial f_2} = \overline{x_3 x_4}\,\overline{x_1 x_2} = 1.$$

Any solution to the latter equation will be a test for the specified fault $f_2 \equiv 1$.

The proposed method of test generation allows one to produce tests for various faults in a circuit [1], multiple faults [27] included, and to propagate the specified fault to a particular circuit output along any preselected path. A limitation of this method is the computational complexity and high storage requirements.

2.4. EQUIVALENT NORMAL FORM (ENF)

Consider a test generation method that makes it possible to find an efficient (nearly minimal) set of tests [21, 28]. The method that is widely known as the equivalent normal form (ENF) is based on finding an ENF for a one-output combinational circuit under test. The procedure consists of the following steps:

(i) All gates in the circuit are numbered consecutively.

(ii) A Boolean expression to be implemented by the circuit in such a way that every jth gate is associated with a pair of subscripted brackets is written. All variables in the Boolean expression are input variables.

(iii) The resulting expression is transformed to disjunctive normal form. The pair of brackets is removed by assigning its subscript to every variable within the brackets. Besides, the redundant terms are not removed.

As an example of constructing an equivalent normal form, let us consider a circuit (Fig. 2.5) for which

$$F = ((x1x2)_1 + (x3x4)_2)_3.$$

By opening the brackets we obtain

$$F = (x1x2)_{13} + (x3x4)_{23}$$

$$F = x1_{13}x2_{13} + x3_{23}x4_{23}.$$

The latter expression is the ENF for the circuit of Fig. 2.5.

In general, the ENF for any digital circuit is characteristic of the fact that any variable of the circuit contains complete information about one of the probable paths in the circuit. A particular path starts in the point of ENF variable application and passes through the gates whose numbers are its subscripts. Thus for the ENF circuit of Fig. 2.5, variable $x1_{13}$ describes a path starting in the application point x_1 and passing through gates 1 and 3.

The test generation algorithm based on the equivalent normal form method consists of the following steps.

(i) The ENF for the circuit under test is constructed.

(ii) A path of the fault under test is defined as a sequence of gates.

(iii) The selected path is specified by an ENF variable whose subscripts are associated with the numbers of the gates in the path.

(iv) Significance is introduced for the selected path. For this purpose, in one of the terms with the selected ENF terms the remaining factors are extended to 1, and in the other terms at least one factor is extended to 0.

As an example of test generation by the under consideration, let us examine the circuit of Fig. 2.5 for $f_2 \equiv 1$. The procedure is as follows:

(i) $F = x1_{13}x2_{13} + x3_{23}x4_{23}$.

(ii) A path through gates 2 and 3 along which $f_2 \equiv 1$ propagates is chosen.

(iii) The path chosen is specified as the variable $x3_{23}$.

(iv) The condition of the substantial nature of this path is ensured.

As a result we obtain $x4 = 1$, $x2 = 1$, $x1 = 0$.

The application of the equivalent normal form method results in an input pattern $x1x2x3x4 = 0101$, which is a test for fault $f_2 \equiv 1$.

A limitation of the method is that its implementation by modern digital circuits containing many gates is extremely laborious.

2.5. DETAILS OF TEST GENERATION FOR SEQUENTIAL CIRCUITS

While the algorithms discussed are efficient in test generation for combinational circuits, the testing of sequential circuits still remains a problem due to the fact that the sequential circuit behaviour depends not only on the present set of input variables but also on the set of past inputs. This makes it necessary to apply a sequence of input patterns to the circuit inputs to detect a single stuck-at fault.

There are two distinct approaches to the problem of generating tests for sequential circuits.

(i) Conversion of a given sequential circuit into a set of identical combinational circuits. In this case most of the methods earlier discussed for combinational circuits can be applied.

(ii) Verification whether or not a given sequential circuit is operating in accordance with its state table.

A general model of a sequential digital circuit is represented by the so-called

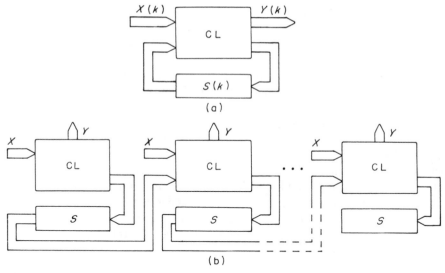

Fig. 2.6. (a) A model of a sequential circuit and (b) its iterative circuit: CL, combinational logic; $X(k)$ and X are inputs; $Y(k)$ and Y are outputs; $S(k)$ and S are states of memory elements; $k = 0, 1, 2, \ldots$

Huffman model (Fig. 2.6(a)) where the values $Y(k)$ are determined by the state $S(k)$ of memory elements (flip-flops) and input values $X(k)$. A successive state $S(k+1)$ of memory elements is also determined by the state $S(k)$ of flip-flops and input variables $X(k)$, where $k = 0, 1, 2, \ldots$.

The given model can be represented as some copies of the same sequential circuit with the state of the first copy being an input for the second, the state of the second copy being an input for the third, etc. As a result of transformation, the initial diagram assumes the form of a combinational network consisting of a sequence of identical circuits whose inputs are input variables of the sequential circuit and the state associated with the preceding circuit. The number of circuits included in the model of the initial network (Fig. 2.6(a)) equals that of its l states.

Such transformation replaces the problem of testing sequential circuits by the problem of testing combinational circuits. However, a single fault in the sequential circuit is represented in this case as an l-multiple fault in its iterative model (Fig. 2.6.(b)). Therefore, with an increase in the number of faults and complexity of sequential circuits, the approach becomes unrealistic. Moreover, a problem occurs that calls for detecting a fault independently of the initial state of the circuit under test, which requires one to derive the so-called homing sequence. More acute is the problem of deriving homing sequences for another approach, which relies on determining the behaviour of a sequential circuit by input and state tables [5]. An input table is a function $Y(k) = \lambda[X(k), S(k)]$ that associates the output value $Y(k)$ with an input value $X(k)$ and the circuit state $S(k)$. A successive state $S(k+1)$ of the circuit is determined by its preceding state $S(k)$ and input value $X(k)$ and is expressed by $S(k+1) = \delta[S(k), X(k)]$.

The basic idea of verifying the sequential circuit operation by its state table is that a sequence pair $\{X(k)\}$, $\{Y(k)\}$ can be found such that the output sequence will be $\{Y(k)\}$ with an input sequence $\{X(k)\}$ if and only if there is no fault. To perform such an experiment it is necessary to design a checking sequence that determines uniquely the initial state $S(0)$ for a sequential circuit.

A method based on the state table verification can be considered as a functional testing method [2]. A limitation of the method is that it requires the description of the behaviour of a given digital circuit by the state table, which is not practicable except for a small portion of circuits with insignificant number of memory elements. Moreover, this method produces excessively long test sequences for circuits in which memory elements constitute a significant portion.

An alternative approach, known as the diagram approach [1], uses an input sequence that allows one to obtain all possible states of an initial circuit. By this approach a large amount of circuit faults can be detected. It is independent of particular fault realizations and models, and allows one to detect transient faults of the circuit. A variation of the state diagram approach is one that decomposes a digital circuit into a set of subcircuits and generates a test sequence [29] for each of them. Experimental results show that test sequences obtained by this approach can detect more than 90% of stuck-at faults.

2.6. *ESTIMATING TEST GENERATION EFFICIENCY*

Test generation is most often based on the assumption that only a single stuck-at type fault is present in the circuit under test. It is valid only for circuits that are tested frequently, when the probability of more than one fault occurring is small. Circuits that are tested for the first time as well as LSI and VLSI chips can contain multiple faults. Some statistical studies have shown then that, to establish reliability of a LSI chip, it must be tested for multiple faults composed of at least eight single faults [1].

At the same time, however, test generation for all possible multiple faults is not feasible. For a k-node digital circuit there can exist $2k$ possible faults and the number of possible stuck-at type faults is $3^k - 1$ [1]. Hence, test generation for all possible faults is impracticable even for small networks. For this reason, we are usually forced to reduce the number of faults to be tested by combining, for example, indistinguishable faults into a single set [5]. By 'indistinguishable faults' we shall mean here such faults for which there is no test to distinguish them ($\equiv 0$ input or output faults for an AND gate and respectively $\equiv 1$ input or output faults for on OR gate). Therefore, when generating a test for an n-input AND (OR) gate only $(n + 2)$ rather than $(2n + 2)$ faults of the gate need to be tested. A systematic approach that reduces the number of faults that have to be tested is based on the idea of fault equivalence classes, i.e. such faults that are covered by a single test set. Thus, all possible stuck-at faults of a combinational tree can be detected by a test sequence for stuck-at faults at its inputs, whereas those of a digital circuit with fan-in branches can be detected by a test for stuck-at fan-in/fan-out faults [5].

A more general approach that results in a significant reduction in the number of faults to be tested uses the concept of prime faults [30].

(i) Stuck-at-1 (stuck-at-0) faults for every AND (OR) gate input if that is an input line in the circuit.

(ii) Stuck-at-1 (stuck-at-0) faults for AND (OR) gate inputs if they are fan-out branch lines.

(iii) Stuck-at-0 (stuck-at-1) faults for AND (OR) gate outputs if all their inputs are input lines of the circuit and there are no fan-out lines.

(iv) Stuck-at-0 (stuck-at-1) faults for a NOT gate that is either an input or a fan-out line in the circuit.

The reduction in the number of faults at test generation simplifies the generation process insignificantly. By running the test checker programs [31], the following relation has been obtained

$$t_g = k_1 L^3$$

where t_g is machine time required to generate tests for a digital circuit employing L gates and k_1 is a proportionality factor.

Another approach that is more cost-effective is the one based on the concept of fault simulation. The principle of the approach is that random patterns of input stimuli are applied to two identical circuits or their models. One of the circuits is used as a reference and the other is successively fed with faults [5]. When the reference circuit output responses differ from those of the faulty circuit, the random pattern is considered to be a test for the specified fault. By successively generating all possible faults we can construct a complete test. Then sequential, parallel or deductive fault simulation is used [5]. The efficiency of the approach, which is often referred to as random search [32], is attributed to the fact that a large test set is used for detection of many faults. Time requirements for the random search approach are evaluated by the relation

$$t_s = k_2 L^2$$

where t_s is machine time required for simulating faults in a circuit of L gates and k_2 is a proportionality factor [31].

However, relations obtained for t_g and t_s closely approximate time requirements for test sequence generation only for 70–80% of detectable faults (Fig. 2.7) [5]. Curves A and B (Fig. 2.7) represent the test coverage (percentage of all possible faults as detectable faults) as a function of test generation time for deterministic and random search approaches. By using the two mentioned approaches together, (curve C), better time estimates can be obtained. In this case, random search is used at the first step

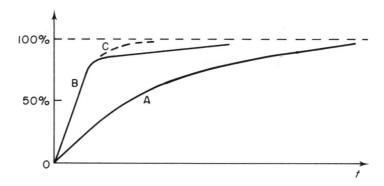

Fig. 2.7. Test coverage versus test generation time: A, deterministic technique; B, random search technique; C, a combination of both

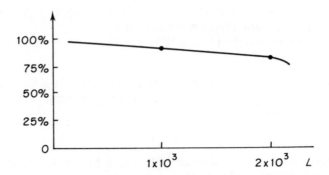

Fig. 2.8. Averaged quality dependence (test coverage) of obtained tests for a digital circuit with L gates

since it allows one to generate a test for faults that are easily detectable by a test pattern set, whereas for faults that are detectable by a specific input pattern one of the directed-search methods (d-algorithm, Boolean difference, etc.) can be used.

Although classic test generation approaches ensure that a high percentage of faults are detected, the problem of efficient test design still remains to be solved, since practically all implemented test generation techniques deal with a class of single stuck-at faults and their associated efficiency estimates are given for the said fault class. As has been noted earlier, actual faults in LSI/VLSI circuits are not exhaustively covered by a single-fault model. These are most commonly short-circuit type faults, programming errors, missing or excess gates in programmable logic arrays (PLAs), etc. [31].

Some papers report on the analysis of the possibility of multiple-fault testing by single-fault test sets. It has been proved [33] that not every multiple fault can be detected by such a test. For certain types of combinational circuit structures, especially those with convergent fan-out lines, more than 75% of multiple faults remain undetectable.

Thus, to account for all real LSI/VLSI faults, we need new approaches to solve the problem of digital circuit testing and diagnostics. Moreover, since the test generation problem belongs to the universal exhaustive search (NP-complete) problems, approximately 10^4 gates is a practical limit to classical test generation efficiency [31]. This is confirmed by an averaged dependence of generated test quality for digital circuits with different numbers of gates shown in Fig. 2.8 (test quality is a percentage of all possible faults as detectable faults in a circuit with L gates) [31].

The use of radically new techniques for solving the problem discussed is caused by the fact that test generation for modern digital circuits is rather time-consuming. A more realistic approach is to account for diagnostic testing problems at the stage of digital circuit design. For this purpose, different logic and engineering design rules that simplify the problem of testing as well as new circuit engineering concepts are used.

Now let us consider major approaches used in circuit design for simplifying the test sequence generation.

3

DETAILS OF TESTABLE DIGITAL CIRCUIT DESIGN

3.1. DESIGN OF TESTABLE COMBINATIONAL CIRCUITS

A number of publications dealing with the design of testable digital circuits normally solve the problem by increasing the number of input and output variables, selecting the components or design procedure. Thus the complexity of circuit implementation increases yet its testing is simplified. One of the first reports on the problem proposed a technique of realizing any arbitrary n-variable Boolean function using AND and EOR gates that perform logic multiplication (conjunction) and modulo-2 addition [34]. The theory of component selection for implementation of function $F(x_1, x_2, \ldots, x_n)$ is based on the Reed–Muller expansion

$$F(x_1, x_2, \ldots, x_n) = C_0 \oplus C_1 \dot{x}_1 \oplus C_2 \dot{x}_2 \oplus \cdots \oplus C_n \dot{x}_n \oplus C_{n+1} \dot{x}_1 \dot{x}_2 \oplus C_{n+2} \dot{x}_1 \dot{x}_3$$

$$\oplus \cdots \oplus C_{2^n - 1} \dot{x}_1 \dot{x}_2 \cdots \dot{x}_n \tag{3.1}$$

where $C_i \in \{0, 1\}$, $i = 0$ to $2^n - 1$, is a constant and $\dot{x}_j \in \{0, 1\}$, $j = 1$ to n, is either a Boolean variable x_j or $\bar{x}_j$ but not both together in the same expression (3.1). An example is a function $F(x_1, x_2, x_3)$ that relies on three variables. By using (3.1) we can write

$$F(x_1, x_2, x_3) = C_0 \oplus C_1 x_1 \oplus C_2 x_2 \oplus C_3 x_3 \oplus C_4 x_1 x_2 \oplus C_5 x_1 x_3$$

$$\oplus C_6 x_2 x_3 \oplus C_7 x_1 x_2 x_3$$

with values $C_i, i = 0$ to 7, being computed by the following expressions [1]:

$$C_0 = F(0, 0, 0)$$
$$C_1 = F(0, 0, 0) \oplus F(1, 0, 0)$$
$$C_2 = F(0, 0, 0) \oplus F(0, 1, 0)$$
$$C_3 = F(0, 0, 0) \oplus F(0, 0, 1)$$
$$C_4 = F(0, 0, 0) \oplus F(0, 1, 0) \oplus F(1, 0, 0) \oplus F(1, 1, 0)$$
$$C_5 = F(0, 0, 0) \oplus F(0, 0, 1) \oplus F(1, 0, 0) \oplus F(1, 0, 1)$$
$$C_6 = F(0, 0, 0) \oplus F(0, 0, 1) \oplus F(0, 1, 0) \oplus F(0, 1, 1)$$
$$C_7 = F(0, 0, 0) \oplus F(0, 0, 1) \oplus F(0, 1, 0) \oplus F(0, 1, 1)$$
$$\oplus F(1, 0, 0) \oplus F(1, 0, 1) \oplus F(1, 1, 0) \oplus F(1, 1, 1).$$

Thus, for example, a function $F(x_1, x_2, x_3) = x_1 x_2 + \bar{x}_1 x_3 + \bar{x}_2 \bar{x}_3$ can be expressed in the Reed–Muller expansion form as

$$F(x_1, x_2, x_3) = 1 \oplus x_2 \oplus x_1 x_2 \oplus x_1 x_3 \oplus x_2 x_3.$$

A hardware implementation of the function is shown in Fig. 3.1.

Digital circuits that have been designed by using the Reed–Muller expansion have the following properties [1].

(i) A circuit implementing a function $F(x_1, x_2, \ldots, x_n)$ whose input nodes are fault-free requires at most $(n + 4)$ tests to detect all single faults.

(ii) If there are faults on all nodes of the circuit, then the test sequence will contain $(n + 4) + n_e$ tests, where n_e is the number of input variables that appear an even number of times in the Reed–Muller expansion (3.1). For the example of Fig. 3.1, $n_e = 2$, since the variables x_1 and x_3 appear twice in the expansion of the function being implemented [1].

In [34] it has been shown that to detect a single fault in a cascade of EOR gates (Fig. 3.1) it is sufficient to apply input sequences that will test each EOR gate for all possible input combinations. For the circuit of Fig. 3.1, such an input sequence is

z	x_1	x_2	x_3
0	0	0	0
0	1	1	1
1	0	0	0
1	1	1	1

In this case the input sequence structure is independent of the number of variables n and contains four tests only. Thus for a five-variable function the sequence would have the form

z	x_1	x_2	x_3	x_4	x_5
0	0	0	0	0	0
0	1	1	1	1	1
1	0	0	0	0	0
1	1	1	1	1	1

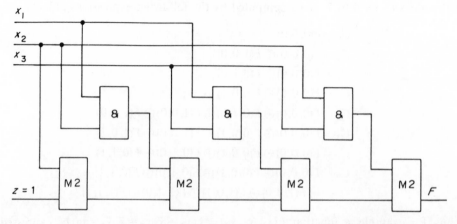

Fig. 3.1. A Boolean function $F(x_1, x_2, x_3) = x_1 x_2 + \bar{x}_1 x_3 + \bar{x}_2 \bar{x}_3$ as the Reed–Muller expansion
$F(x_1, x_2, x_3) = 1 \oplus x_2 \oplus x_1 x_2 \oplus x_1 x_3 \oplus x_2 x_3$

Apart from the four tests used for detecting EOR gate faults, it is necessary to use n tests for an AND gate with $n_e = 0$ [1]. A complete test sequence can be obtained by uniting two sets of tests.

A major limitation of the technique discussed is the increased complexity of circuit implementation as well as the increased number of logic levels (depth) in a circuit, which affects its performance [1]. That is why a number of design techniques have been proposed for the implementation of testable combinational circuits of minimal depth. One of them is a three-level OR–AND–OR design method [35], which results in testable combinational circuits implemented as a three-level structure. The major drawback of the method is restriction on the class of functions that can be implemented.

As mentioned above, testability of a digital circuit is directly related to its observability and controllability. Additional gates inserted in a digital circuit to increase the number of its inputs and outputs allow one to improve these properties and to simplify testing and diagnosing. It has been proved, in particular, that, by inserting additional logic, a digital circuit can be modified so that test sequences of five tests are sufficient to detect all possible faults in the circuit [36]. The design procedure for such a digital circuit consists of the following steps.

(i) Implement the specified Boolean function by using two-input NAND and NOT gates.

(ii) Replace all inverters by two-input EOR gates whose second input will be an additional input in the circuit to which a 1 is normally applied.

(iii) Insert two-input EOR gates in all other inputs of two-input NAND gates and apply a 0 to their second inputs, these inputs being additional.

To examine the application of the discussed procedure, let us use a Boolean function

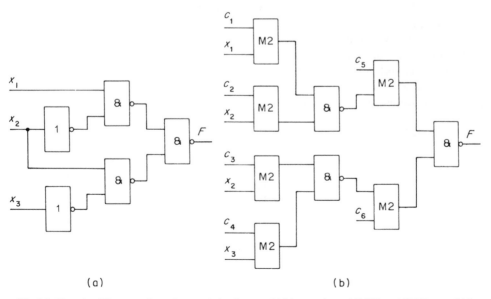

(a)   (b)

Fig. 3.2. Function $F(x_1, x_2, x_3) = x_1 \bar{x}_2 + x_2 \bar{x}_3$ implemented (a) by two-input NAND and NOT gates (b) its modification with two-input OR gates replacing the inverters and two-inputs OR gates connected to the remaining inputs

$F(x_1, x_2, x_3) = x_1 \bar{x}_2 + x_2 \bar{x}_3$ [1] whose implementation by two-input NAND and NOT gates is shown in Fig. 3.2(a). Replacing all inverters by two-input EOR gates and inserting two-input EOR gates to all the remaining input lines of two-input NAND gates, we obtain a modified circuit with six additional inputs c_1, c_2, c_3, c_4, c_5, and c_6 (Fig. 3.2(b)). With the values of variables on additional input lines $c_1 = 0, c_2 = 1, c_3 = 0$, $c_4 = 1, c_5 = 0$ and $c_6 = 0$, the circuit under consideration implements the specified function $F(x_1, x_2, x_3) = x_1 \bar{x}_2 + x_2 \bar{x}_3$.

Let us first consider the basic module (Fig. 3.3) used to generate test sequences for a circuit. All four input combinations are required to detect all possible faults for a two-input EOR gate. The two-input EOR gate outputs p_1 and p_2 will be input tests for the two-input NAND gate and its output responses will be applied to the OR gate input in a successive module. Therefore the combination $p_1 = 1$ and $p_2 = 1$ must be applied twice to the NAND gate inputs to ensure that a 0 will appear on its output. Thus complete testing of each module requires five tests

x_1	c_1	x_2	c_2	p_1	p_2	f
0	0	0	0	0	0	1
0	1	0	1	1	1	0
1	1	1	0	0	1	1
1	0	1	1	1	0	1
1	0	1	0	1	1	0

where p_1, p_2 and f are responses to sets $x_1 c_1 x_2 c_2$.

The analysis of test sets shows that for complete testing of a modified circuit each five-bit test sequence on the primary and NAND gate inputs must belong to the set $\{X_0, X_1, ..., X_9\}$ [1], where $X_0 = 00111$, $X_1 = 01011$, $X_2 = 01101$, $X_3 = 01110$, $X_4 = 10011$, $X_5 = 10101$, $X_6 = 10110$, $X_7 = 11001$, $X_8 = 11010$ and $X_9 = 11100$. Now, the test sequence generation consists of assigning input sequences that belong to the given set. Then we derive input sequences for additional inputs in a circuit (two-input OR inputs) to produce sequences belonging to the set $\{X_0, X_1, ..., X_9\}$ on the two-input NAND gate inputs such that their application to the NAND gate does produce a sequence belonging to the same set. The procedure is repeated until an output sequence produced for the circuit is one that cannot belong to the specified set.

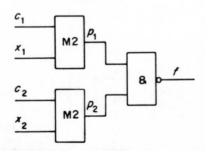

Fig. 3.3. Basic module for constructing a two-input NAND and OR circuit

Applying the above procedure to the circuit of Fig. 3.2(b) the following test set

x_1	x_2	x_3	c_1	c_2	c_3	c_4	c_5	c_6
0	1	1	1	0	0	1	1	0
1	1	0	0	1	1	0	0	0
1	0	1	1	1	1	0	0	1
1	1	0	0	0	0	1	0	0
0	0	1	0	0	0	0	1	1

consisting of only five input tests is obtained [1].

By applying the alternative combinational circuit design technique [37] based on the principle that any n-input AND, OR, NAND and NOR gate can be tested by $(n+1)$ tests, we can obtain better estimates. Then all possible single and multiple stuck-at faults can be detected. Hence, if a gate contains only two inputs it can be tested by three tests. This statement is also valid for a digital circuit that has been realized using two-input gates by the technique of [37]. For example, a three-input OR gate can be replaced by a three-level network consisting of two-input OR, AND and OR gates (Fig. 3.4). When operated, the second input c_1 of the AND gate is applied with a logic 1. This circuit can be tested for all possible faults by using a sequence of three input tests

x_1	x_2	x_3	c_1
0	0	0	1
0	1	1	0
1	0	0	1

Despite the advantages of the techniques discussed for testable combinational circuit design, they are impracticable due to hardware redundancy of designed circuits and the increase in number of inputs and outputs. They are especially inadequate for LSI and VLSI. It has been proved that major difficulties arise at designing tests for sequential circuits employed in the majority of currently used digital networks rather than for combinational circuits.

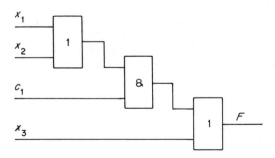

Fig. 3.4. A three-input OR gate represented as a three-level circuit employing two-input OR, AND and OR gates

3.2. SCAN TECHNIQUES

The testing of sequential circuits is complicated because of difficulties in setting and checking the states of memory elements that form the majority of such circuits. Therefore, remembering that test pattern generation for complex sequential circuits (see Sec. 2.5) is impracticable, we must modify a circuit so as to simplify its testing. The simplest way of modifying a sequential circuit is to divide it into two components, combinational and memorizing, to be further tested individually.

One of the first implementations of the above approach is the scan-path technique, which is often referred to as the shift-register-modification approach [38]. The basic idea behind it is to incorporate redundancy with the aim of connecting up memory elements into a shift-register circuit. In the general case, the modified sequential circuit will comprise a combinational circuit (CC), memory elements $s_1, s_2..., s_l$, input and output lines $x_1, x_2,..., x_n$ and $y_1, y_2,..., y_m$ respectively, system clock input C_1, and shift-register mode input C_2, shift-register input D, its output Q and l two-input multiplexers.

By incorporating extra multiplexers in a sequential circuit we can improve its testability.

(i) Memory elements can be tested isolated from the rest of the circuit.

(ii) Any successive state of the circuit can be set independent of its current state.

(iii) Output values of the combinational portion applied to the memory elements can be observed directly.

The above steps are performed as a result of setting up a shift register of memory elements with a memory element output being connected via the multiplexer to the input of another memory element. The specified operation is implemented by applying a control signal C_2.

The resulting shift-register input is D and its output is Q. When an opposite value of C_2 is applied, the multiplexers connect the outputs of the combinational portion of the circuit to the inputs of memory elements and the circuit, having returned to its initial configuration, can be used as intended.

The procedure of testing a circuit that employs the elements used to form a shift register consists of the following steps [14].

(i) Set up a shift register of memory elements in a circuit by specifying $C_2 = 1$.

(ii) Test the status and operation of each memory element in a shift register:

(a) All memory elements are initialized to 0 by applying the sequence $000 \cdots 0$ to input D and the sequence of clocks that is greater or equal to l to input C_1.

(b) The sequence $100 \cdots 0$ is applied to input D and clocks are applied to input C_1. As a result each memory element is successively set to 1.

(c) Steps (a) and (b) are repeated with a single 0 flushed through a background of 1 s. This test for output responses at shift-register output Q checks the ability of each memory element to assume a steady-state 0 and a steady-state 1:

(d) Shift test is performed by applying the sequence $00110011 \cdots$ to input D. In this case each memory element flushes through all combinations of present state and future state.

(iii) The shift register is initialized to 0.

(iv) Specify the test pattern by applying input variables to the CC inputs and storing variables applied to the CC from memory elements at the shift register.

(v) The value C_2 is set to 0 causing the circuit to return to its normal configuration.

(vi) The response to the test pattern is observable at the outputs that are the outputs of the circuit $(y_1, y_2, ..., y_m)$. A clock C_1 is generated to store the responses of the CC outputs connected to the memory elements.

(vii) Specify $C_2 = 1$ to form a shift register.

(viii) By successively applying pulses C_1, observe the responses of CC outputs loaded to memory elements.

(ix) Such a procedure is repeated for any successive test pattern starting from step (iv).

The most prominent advantage of the discussed approach is that a sequential circuit is converted into a combinational circuit at testing, thus making test generation for the circuit easier than for the sequential circuit. However, the scan-path technique has some disadvantages: the first is the necessity of using bistable memory elements to prevent race conditions that may cause malfunctions [38]; the second is the significant time expenditure required for testing that is determined by sequential shift-in and shift-out modes. The shift-register length is a major value characterizing the effect required for testing a circuit.

By dividing an original shift register into some short registers it is possible to make testing less time-consuming. In such a case, the time required for storing test data or reading the CC response is determined by the length of the longest register among them. However, this technique can greatly increase the number of primary inputs and outputs, which may prove to be a major disadvantage in LSI and VLSI design.

A random-access scan technique is an alternative approach to the scan-path technique [39]. By using the random-access scan approach a sequential digital circuit with randomly addressed memory elements is designed, thereby allowing any memory

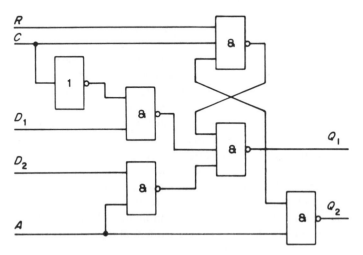

Fig. 3.5. Functional diagram of an addressable memory element: R, reset; C, clock; D_1, data input; D_2, additional data input; A, address line; Q_1, a 1 at memory element output; Q_2, an additional data output

element state to be set, reset and checked independently. Fig. 3.5 shows a typical structure of an addressable storage element. It has two extra inputs D_2 and A, a data input and an address line, as well as an extra data output Q_2. Inputs D_2 and outputs Q_2 from all memory elements are ORed together. Such a digital circuit will have only one extra input and one output. The required memory element is selected by enabling the address line A in only one memory element. For this purpose, the configuration of a circuit can comprise a decoder and an address register, which are optional in the general case. Thus if a circuit is not critical to additional inputs, it can have no address register and specify the address of the selected memory element at its external inputs. Inputs R, C, D_1 and Q_1 serve as reset, clock, data inputs and a flip-flop output (Fig. 3.5). They are used for their designated purpose.

The random scanning technique discussed above requires three or four additional gates per memory element as well as additional addressing circuits [31]. The same as for the scan-path technique, circuit redundancy and the number of additional inputs are determined by the number of memory elements in the circuit being implemented.

3.3. LEVEL-SENSITIVE SCAN DESIGN (LSSD)

One of the best known methods for synthesizing testable digital circuits is the level-sensitive scan design (LSSD) [40, 41]. It is a logical follow-on of the scan-path technique and is based on two concepts [40, 42].

(i) Any change in a digital circuit state involves the changing of synchronization level rather than its edge (level-sensitive).

(ii) At testing, all memory elements in the circuit are interconnected into a shift register to which any circuit state can be written to be used for analysing the response of its combinational part (scan design).

To implement the method, we need a memory element that is a two-bit shift register employing two level-sensitive latches L_1 and L_2; latch L_2 is used only for diagnostic testing, and latch L_1 is used both for one-line operation and for testing procedure [40].

Latch L_2 has a single data input that is connected to the output of latch L_1 and the clock input B that is used for loading L_1 data into L_2. The output Q_2 of latch L_2 in one of the memory elements is connected to input D_2 of L_1 in another memory element. Thereby all memory elements are interconnected into a shift register, which operates when clock signals A and B are applied to its inputs. Inputs A and B in all latches are interconnected whereas clocks are applied sequentially first to input A and, when clock A changes, to input B. Inputs D_1 and C as well as outputs $Q_1\bar{Q}_1$, which are data and clock inputs and data outputs, respectively, are used as specified by the function implemented by the digital circuit.

The network of Fig. 3.6, which contains combinational circuits CC1 and CC2 and two memory elements, is an example of a sequential digital circuit implemented by the level-sensitive scan design technique.

In the normal mode of operation, the network is fed with inputs $x_1, x_2, \ldots, x_n$ and a clock signal C, whereas the values of $y_1, y_2, \ldots, y_m$ are produced on its outputs. In the test mode, the test-data inputs (TDI) are fed with test patterns that are shifted

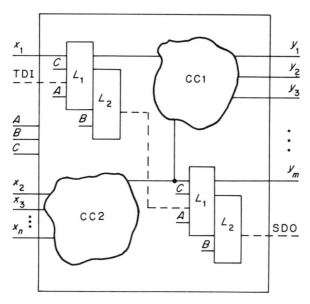

Fig. 3.6. An implementation of a sequential circuit by the LSSD technique

in by clock signals to inputs A and B. The response of the network's combinational section can be obtained at its sequential data output (SDO).

In spite of its simplicity, the technique used for designing digital devices required a specific set of design rules to be obeyed [43].

(i) All memory elements in a sequential circuit must be implemented as two-place shift registers employing level-gated latches L_1 and L_2.

(ii) Memory elements must be controlled by two or more non-overlapping clock sequences. This is required because the two memory elements that are connected in series change their state at different times.

(iii) Neither of the memory elements can be controlled by a clock derived from its output signal.

(iv) All clock inputs must be controlled independently such that it is possible to set any of the clock inputs to the ON state when all the remaining inputs are OFF.

(v) Clock primary inputs may be connected only to clock inputs of memory elements.

(vi) All memory elements must be interconnected into a single shift register whose primary input and output are controllable by shift clocks at inputs A and B. Memory elements may also be interconnected into several shift registers.

(vii) All clock sequences other than those produced at inputs A and B must be OFF at testing.

(viii) Any clock produced at input A must be time-independent of that produced at input B and vice versa.

Only a digital circuit designed in compliance with these rules can meet the requirements of the LSSD technique. It should have the following properties.

(i) Any digital circuit implemented by the LSSD technique requires one to design testability in its combinational section only. Therefore any test design approaches widely used for combinational circuits in recent years are applicable.

(ii) An LSSD circuit operates correctly irrespective of such characteristics as pulse edge rise time and fall time, races, etc.

The use of digital circuits implemented by LSSD for designing digital devices and systems allows one to enforce unified diagnostic testing at all levels: a chip, a module comprising lots of chips, and the whole device or system. It should be noted, however, that the advantages of the LSSD technique discussed above are achieved with the hardware overhead of 20% and the use of four additional primary inputs/outputs in each circuit. Besides, an LSSD circuit exhibits some reduction in its speed due to its synchronous operation defined by the design rules.

The LSSD technique has proved that by applying the following rules at the design stage we can make digital circuits controllable. Let us consider some general rules, which, when applied at the circuit design stage, will greatly simplify test pattern generation as well as testing and diagnosing of the circuits.

3.4. PRACTICAL RULES OF DESIGNING TESTABILITY TO DIGITAL CIRCUITS

In practice, the principal emphasis has been on the generation of fault simulation and test programs for field-replaceable units (FRUs) in digital devices. This may be explained by their relatively low functionality, which allows one to detect a high percentage of faults at the stage of manufacturing supervision. To make the procedure of FRU testing and diagnosing easier, various design rules and limitations are applied to ensure 90–100% fault coverage and fault isolation down to three or four integrated circuits (ICs) for the majority of FRUs by the fault detection tests generated for the class of stuck-at faults [5, 31].

In [44] are presented the rules that have been used for designing testability to digital circuits. They can be classified into two categories: ease of test generation and ease of circuit test and diagnosis. Let us successively discuss the most important rules and limitations from both categories.

(i) The first rule requires one to maximize the controllability and observability of the digital circuit. A high controllability level is achieved due to the fact that any node of a digital circuit can be easily set to any desired level (0 or 1). High observability is achieved when the state of any node in the circuit can be easily defined by its external nodes. When both controllability and observability values are reasonable, then it follows that any circuit element is easily controlled and observed. This allows one to generate efficient test sequences for fault detection in a circuit.

When designing a digital circuit it is necessary to provide for good controllability of such nodes as inputs to flip-flops, counters, set/reset registers, clocks, address inputs to multiplexers and demultiplexers, read/write inputs to memory elements, etc. Also, provision should be made for good observability on the following nodes of the circuit: control lines, outputs from flip-flops, counters and registers, outputs of such data-transmitting units as multiplexers, demultiplexers, coders, decoders, etc. The above listed nodes of digital circuits are commonly called test points, which are also determined by controllability/observability measurement techniques (see Sec. 1.2). Low controllability or observability on any node in a digital circuit points to the fact that the node belongs to the set of test points.

Once all the test points are known, we may modify the basic design to improve the control and observation features. For this purpose, the following methods may be used:

(a) introduction of extra gates,

(b) provision of additional external inputs and outputs, and

(c) use of logic and mechanical adapters, which allow one to break the internal connections and make them accessible in the circuit.

The example in Fig. 3.7 illustrates the use of the above methods; the circuit has been provided with an extra external inputs and output as well as with an AND gate for improving observability/controllability on its node f.

All successive rules, although specific in their nature, are also directed to some extent to the improvement of the basic design from its testability standpoint, i.e. controllability and observability.

(ii) The second rule is aimed at avoiding logical redundancy of the circuit. A digital

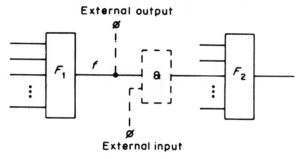

Fig. 3.7. An example of improving observability/controllability by introducing an extra two-input AND, external input and output to the digital circuit

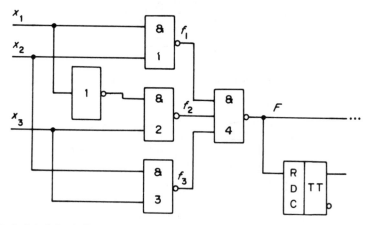

Fig. 3.8. A digital circuit illustrating the effect of a redundant node on fault detection procedure

circuit node is considered logically redundant if all output values of the circuit are independent of the variable produced on the node for all possible input stimuli [14]. An exmple of a logically redundant node is the output f_3 from the two-input NAND gate in the circuit which implements the function $F(x_1, x_2, x_3) = x_1 x_2 + \bar{x}_1 x_3 + x_2 x_3 = x_1 x_2 + \bar{x}_1 x_3$ (Fig. 3.8). In fact, the value of $F(x_1, x_2, x_3)$ is determined by only the values $f_1 = x_1 x_2$ and $f_2 = \bar{x}_1 x_3$ and is independent of $f_3 = x_2 x_3$.

The presence of logically redundant nodes may cause racing or fault masking on non-redundant nodes in the circuit. Thus (Fig. 3.8) with fault $f_3 \equiv 1$ present and the change of variable x_1 from high state to low state, for $x_1 = 1$ and $x_3 = 1$ an intermittent negative pulse may occur causing the D-type flip-flop to operate.

A fault on a redundant node when undetectable may mask subsequent detection of a second fault on a non-redundant node. Thus (Fig. 3.9) a stuck-at-1 fault ($\equiv 1$) at the third input to the first three-input NAND gate is undetectable. Hence the fault $\equiv 0$ at the second input to the second two-input NAND gate will also be undetectable. However, when absent, the input pattern $x_1 = 1$, $x_2 = 1$, $x_3 = 0$ will be used as a test for $\equiv 0$ at the input to the two-input NAND gate.

(iii) The third rule is to provide for breaking feedback loops. Uncontrollable feedback loops are the source of some problems at test generation and fault simulation. An example of such a loop may be a fragment of circuit shown in Fig. 3.10, where the presence of 1s at the secondary inputs to two-input NAND gates causes

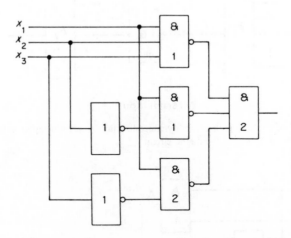

Fig. 3.9. A digital circuit illustrating the redundant node effect of masking the faults on its other nodes

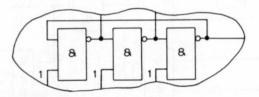

Fig. 3.10. A fragment of a digital circuit that illustrates the effect of uncontrollable feedback loops on test generation and fault simulation

high-frequency oscillations, which initiate false operation of memory elements in the circuit.

(iv) When designing complex digital circuits with many gates L, provision should be made for partitioning them into smaller subcircuits to reduce test generation effort. Thus if the circuit containing L devices can be partitioned into two subcircuits with $L/2$ devices each, the total test design effort is reduced by a factor $k_1L^3/(k_1L^3/8 + k_1L^3/8) = 4$ (see Sect. 2.6).

(v) All unused inputs to logic devices in a digital circuit must be terminated to the power supply output or the zero-level bus through a suitable resistor.

(vi) All clocks used in a circuit design must be controllable and testable by external outputs of the circuit. For that purpose, the clock output must be terminated to an external pin and must have an externally controlled circuit for throwing off the load [31].

(vii) The digital and analogue units in a digital circuit must be kept physically apart. This rule may be explained by the fact that the strategy of analogue circuit testing is substantially different from that for testing and diagnosing digital circuits.

The discussion of design rules used for simplifying the test generation procedure may be continued with an account of specific features of components, structural considerations, layout rules, etc. This is attributed to a rather wide range of factors affecting the test generation problems.

The rules relating to the problem of simplifying the tests and diagnostic procedures are as important as the rules discussed above. Let us consider some of them [14, 31].

(i) Avoid the use of wired AND or wired OR gates for elements with high fan-in

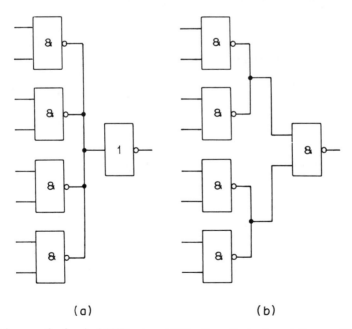

(a)   (b)

Fig. 3.11. (a) An example of a wired AND gate and (b) its alternative implementation providing for deeper fault localization

lines, because it prevents one from localizing a fault down to the gate level. Therefore it is more practicable to use wired AND or wired OR gates for small fan-in only. From this standpoint, the circuit of Fig. 3.11(b) turns out to be more preferable than that of Fig. 3.11(a), which implements the same Boolean function. Therefore diagnostic ambiguity will be lower at fault localization [31]. Similarly, the circuit with high fan-out lines (Fig. 3.12(a)) should be designed the same as that of Fig. 3.12(b).

(ii) Implement the digital circuit so as to locate equivalent faults to the same integrated circuit. In this case by 'equivalent faults' we shall understand the digital circuit faults that are detectable by the same sequence of test patterns. An example of such faults are $x_1 \equiv 1$, $x_3 \equiv 0$ and $f_2 \equiv 1$ of the circuit shown in Fig. 3.13(a). All the listed faults may be located by the same test pattern $x_1 x_2 x_3 = 011$.

When implementing the above circuit, it is preferable to use an integrated circuit with two-input NAND gates (Fig. 3.13(b)) rather than an inverter from another integrated circuit (Fig. 3.13(c)). In the first case, the fault is located down to one IC, whereas in the second case it is located down to two ICs.

(iii) Provide for breaking long counters up into a chain of short counters. This may be explained by the fact that for testing a 16-bit counter fully, $2^{16} + 1 = 65\,537$ input pulses should be applied. At the same time, only $2(2^8 + 1) = 514$ pulses should be applied for testing two 8-bit counters. The long counter may be partitioned the same as shown in Fig. 3.14. In this circuit, enabling pulses are applied to the secondary inputs 2 and 3 to the two-point NAND gates in the normal count mode. Hence the circuit functions as a $2n$-bit counter. However, with a disabling pulse applied to input 2 to the first two-input NAND gate, the counter transforms to a circuit comprising two independent n-bit counters whose inputs are 1 and 3, respectively (Fig. 3.14).

(iv) When implementing a digital circuit, avoid the use of customized components

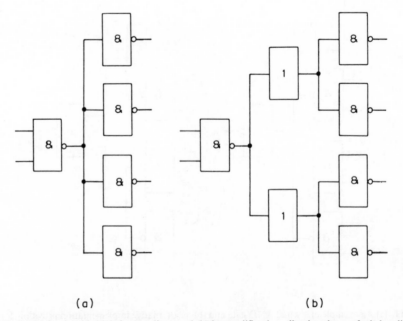

(a)   (b)

Fig. 3.12. (a) A circuit with high fan-out lines and (b) its modification allowing deeper fault localization

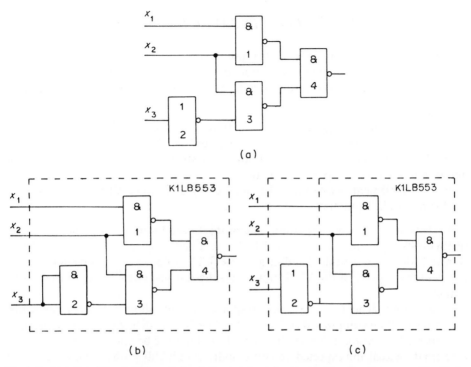

Fig. 3.13. (a) A digital circuit with equivalent faults and circuit implementation that allow one to identify the fault (b) down to the integrated circuit and (c) down to two integrated circuits

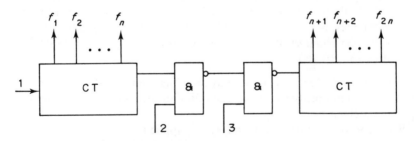

Fig. 3.14. A 2n-bit counter partitioned into two n-bit counters: 1 and 3, data inputs; 2, control input

that may be incompatible as to voltage levels associated with logic variables. The violation of this rule may complicate the requirements of testing equipment.

The design rules listed above, when obeyed, allow one to simplify the circuit testing procedure, which implies the use of complex test equipment, for applying input test sequences and analysing output responses. The application of the self-test concept to digital circuits evolves from the ideas behind the solution of digital circuit test and diagnosis problems.

3.5. DESIGN OF SELF-TESTABLE DIGITAL CIRCUITS

The idea of self-testable digital circuits is used to enhance testability of LSI and VLSI circuits and is based on the application of built-in equipment for test sequence generation and test response evaluation. Self-testing is based on the following principles [1].

(i) Test sequences are generated immediately on the VLSI chip.

(ii) Responses to test stimuli are also stored on the VLSI chip.

(iii) The only operations required for self-testing of the VLSI chip are test initialization and test response evaluation.

(iv) To implement self-testing, the number of additional VLSI chip pins and hardware should be minimal.

An elementary built-in test pattern generator is a pseudorandom sequence generator (PRSG) [45,46], which is configured by using a shift register existing in the VLSI chip and additional modulo-2 adding gates [47]. The PRSG generates pseudorandom test patterns applied to the VLSI inputs and the test responses are produced on its outputs.

Various approaches for compressing response data into short signatures (keywords) are used to obtain compressed self-test data [46,47]. Any difference in the actual and the expected signature indicates that the VLSI chip may be faulty whereas a match between the actual and expected signatures indicates a highly probable good condition of the chip.

A self-test approach using signature analysers based on polynomial division as the facility for test response compression is currently being widely investigated in the literature [47, 48].

In a majority of cases, the BILBO (Built-In Logic Block Observation) structure [49, 50], which allows one to realize a number of functions required for the use of self-test techniques, is examined. Fig. 3.15 shows the logic diagram of a BILBO for a generating polynomial $\varphi(x) = 1 \oplus x^3 \oplus x^4$. Consider the basic functions that can be performed by the BILBO.

(i) Every memory element is set to 0 by applying control signals $c_3 = 0$ and $c_2 = 1$. Under these conditions, the input to each flip-flop becomes a logic 0 that is stored by applying a clock to input c_1.

(ii) With control signals $c_2 = 1$ and $c_3 = 1$ applied, the memory elements in the discussed structure operate as expected in the VLSI chip under consideration. Each memory element can be loaded with input data applied to inputs x_i, $i = 1$ to 4, by a clock c_1 and unloaded through the y_i outputs.

(iii) For conditions $c_2 = 0$, $c_3 = 0$ and $c_4 = 1$, the structure under consideration is converted into a pseudorandom test pattern generator whose equivalent circuit is shown in Fig. 3.16(a). The resulting structure allows one to generate a pseudorandom test pattern applied to a combinational portion of the VLSI chip [47]. The properties of generated test patterns are discussed in more detail in subsequent chapters.

(iv) By inverting control signal c_3, the circuit is converted into a multiple-input signature analyser whose input patterns will be given by x_i, $i = 1$ to 4. Compression of the above patterns will result in a signature (the contents of memory elements in a

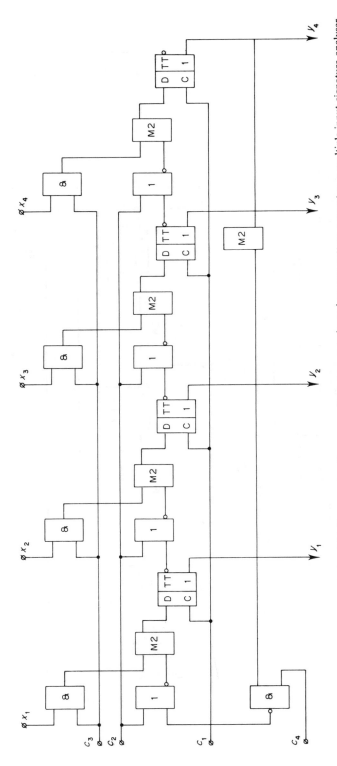

Fig. 3.15. Logic diagram of a BILBO that initializes to zero all memory elements, a pseudorandom test pattern generator, a multiple-input signature analyser and memory elements: x_1, x_2, x_3, x_4 are data inputs; y_1, y_2, y_3, y_4 are data outputs; c_1, c_2, c_3, c_4 are control inputs

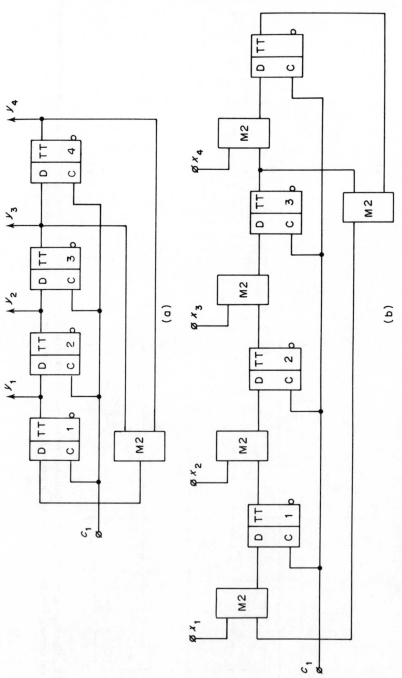

Fig. 3.16. Logic diagrams of (a) pseudorandom test pattern generator and (b) signature analyser

structure of Fig. 3.16(b)). The properties of signatures obtained in such a way as well as the validity of testing and diagnosing by them will be discussed below.

The BILBO technique is discussed in more detail in [1, 14], which analyse its efficiency for self-testing of LSI and VLSI chips designed for scanning the states of memory elements.

The use of PRSG for implementing self-testable LSI/VLSI chips allows one to eliminate the time-consuming procedure of test pattern generation. In this case, however, the problem of test response validity still remains unsolved whereas the testing time increases substantially. Because of this, exhaustive testing techniques [51, 52] based on the concept of generating all possible test patterns to be applied to a circuit under test in the process have recently been widely discussed.

The exhaustive testing techniques are based on the following two concepts:

(i) A VLSI structure is partitioned into subcircuits so each subcircuit has few enough inputs that all possible combinations can be generated.

(ii) Additional gates are inserted to ensure the possibility of analysing the responses of each subcircuit.

An illustration of exhaustive testing implementation is the diagram of Fig. 3.17. It is partitioned into two subcircuits with four input lines each. For analysing each subcircuit response, two-input OR gates and control inputs z_1, z_2 are added.

With $z_1 = 0$ and $z_2 = 0$, a modified diagram of Fig. 3.17 behaves the same as the original one. With $z_1 = 0$ and $z_2 = 1$, the first subcircuit is tested by applying all possible combinations of variables $x_1 x_2 x_3 x_4$. The second subcircuit is tested similarly. The resulting total time required for the modified circuit verification testing will be one-eighth as large as that required for the original circuit testing.

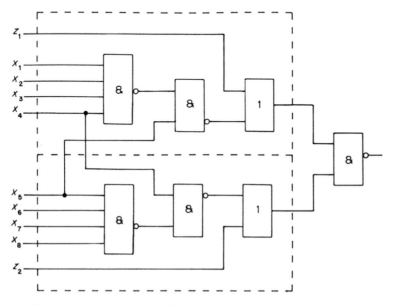

Fig. 3.17. A digital circuit partitioned into subcircuits for exhaustive testing

In a more general case, partitioning an original circuit into multiple subcircuits suggests the use of extra logic for applying test patterns to the lines that interconnect the subcircuits and the analysis of resulting responses. A method based on the use of multiplexers for the purpose has been presented in [1].

The use of built-in test pattern generators, response analysers as well as circuit engineering techniques based on circuit partitioning into subcircuits allows the problems of testing to be solved completely even at the design stage. Then the circuit being tested will generate test patterns, analyse responses and signal its status by itself.

It is evident that implementation of self-testing LSI/VLSI chips requires the solution of such major problems as design of test pattern generators and response analysers, since test coverage depends on the properties of these units. Therefore, let us successively discuss the design concepts for test pattern generators and output response compressing circuits.

4

DESIGN OF CIRCUITS TESTABLE BY COMPRESSED ESTIMATION OF THEIR RESPONSES

4.1. CLASSIFICATION OF OUTPUT RESPONSE COMPRESSION TECHNIQUES FOR DIGITAL CIRCUITS

A classical strategy of digital testing is based on generation of test sequences that allow one to detect the specified sets of circuit faults. As a rule, the test sequences as well as the reference output responses of the circuits to these sequences must be stored for testing. In the course of testing, the actual output responses compared with the reference values allow one to judge the state of the circuit under test. When the resulting circuit responses and the reference values are identical, the circuit is certified to be fault-free; otherwise, the circuit has a fault and is considered to be bad.

For the majority of circuits that are currently manufactured, the classical approach is time-consuming in terms of test-pattern generation and test application (see Sect. 2.6). Moreover, since the amount of test data and reference output responses is large, complex hardware is required for carrying out a test experiment. Therefore, the cost and time requirements for implementation of the classical approach grow faster than the complexity of circuits utilizing it.

That is why new approaches that can significantly simplify both test generation and test application have been proposed. In the general case, the proposed approach implementation can be illustrated by a network whose basic functional units are a test sequence generator and a data compressor [53].

To avoid complex design of test sequences, the techniques selected for test generator implementation must be very simple. Among them are the following algorithms.

(i) Generation of all possible input test patterns, i.e. exhaustive testing for binary combinations. This algorithm produces the so-called count sequences.

(ii) Generation of random test patterns with the desired probability of a 1 or 0 to appear at each input to a digital circuit.

(iii) Generation of pseudorandom test sequences.

The key feature of the above algorithms is that their application for test sequence generation produces a very long sequence. Therefore, the length of responses obtained on the outputs of the circuit under test is identical. Although the problem of storage does not exist for test generators that produce count, random and pseudorandom sequences, it does exist for output responses of each circuit. To reduce the volume of reference output data we may simply obtain an integral estimation of lesser size. For this purpose data compression algorithms are used [54–56]. Owing to their application, compact estimations of compressed data are produced. These estimations are commonly called checksums, keywords, syndromes or signatures for the appropriate nodes of a digital circuit for which one of the data compression algorithms is used. Consider data compression algorithms for a binary sequence $\{y(k)\}$ consisting of l successive binary variables

$$y(1)y(2)y(3)\cdots y(l).$$

One of the first algorithms for obtaining compact estimations is the transition count algorithm given by [57]

$$R_1 = \sum_{k=2}^{l} [y(k-1)\oplus y(k)] \tag{4.1}$$

where R_1 is the number of value changes in the sequence $\{y(k)\}$. Thus, for example, $R_1 = (1\oplus 0) + (0\oplus 1) + (1\oplus 1) + (1\oplus 0) = 3$ for $y(1)y(2)y(3)y(4)y(5) = 10110$.

Modifications of the transition count algorithm are the following equations [57]

$$R_2 = \sum_{k=2}^{l} \overline{y(k-1)\oplus y(k)}$$

$$R_3 = \sum_{k=2}^{l} \overline{y(k-1)}y(k)$$

$$R_4 = \sum_{k=2}^{l} y(k-1)\overline{y(k)}$$

where R_2 is the number of unchanged values of symbols in the sequence $\{y(k)\}$. By using (4.1), we obtain $R_2 = l - R_1 - 1$ (R_3 and R_4 are respectively the number of changes between 0 and 1). Thus, for the above example, $R_3 = \overline{1}\cdot 0 + \overline{0}\cdot 1 + \overline{1}\cdot 1 + \overline{1}\cdot 0 = 1$ and $R_4 = 1\cdot\overline{0} + 0\cdot\overline{1} + 1\cdot\overline{1} + 1\cdot\overline{0} = 2$. The sum $R_3 + R_4$ equals R_1 and $l - R_2 - 1$.

Another technique, i.e. ones counting, is based on the use of the relation

$$R_5 = \sum_{k=1}^{l} y(k) \tag{4.2}$$

where R_5 is the number of 1s in the sequence $\{y(k)\}$. For the sum of zero values R_6 determined by the relationship

$$R_6 = \sum_{k=1}^{l} \overline{y(k)}$$

the condition $R_6 = l - R_5$ is satisfied. This technique is most widely used for syndrome testing [58–60] based on the analysis of R_5.

The algorithms discussed for generating compact estimations from output responses suggest the use of deterministic test generation techniques.

For a random test sequence generator, the probability $P[y(k) = 1]$ of a 1 to appear in an output sequence is based as compact estimate R_7 [61–64]:

$$R_7 = P[y(k) = 1]. \tag{4.3}$$

The use of pseudorandom test sequences normally suggests that signature analysis serves as a compression technique for digital circuit responses [65]. Signature

$$R_8 = a_1(l)a_2(l) \cdots a_m(l)$$

is generated by the algorithm

$$a_1(0) = a_2(0) = \cdots = a_m(0) = 0$$

$$a_1(k) = y(k) \oplus \sum_{i=1}^{m} \alpha_i a_i(k-1) \tag{4.4}$$

$$a_j(k) = a_{j-1}(k-1) \qquad j = 2 \text{ to } m, k = 1, 2, 3, \ldots$$

where $\alpha_i \in \{0, 1\}$, $i = 1$ to m, are determined by a generating polynomial $\varphi(x) = 1 \oplus \alpha_1 x^1 \oplus \alpha_2 x^2 \oplus \cdots \oplus \alpha_m x^m$ used for implementation of a signature analyser [65].

Any of the above techniques for compact estimations allows one to reduce the volume of data to be compressed by a factor of

$$l/\text{int}\,(\log_2 l)$$

on average where int (l) is the nearest integer that is greater than or equal to l. Thus, for the sequence $\{y(k)\}$ consisting of $2^{10} = 1024$ data bits, the compact estimate will be 11 bits long.

A major advantage of data compression techniques is that the volume of data to be stored for test application is significantly reduced. It should be noted, however, that similar integral estimates may result from compression of different sequences, thereby implying that test result distinguishability is degraded. In practice, development of a test method requires that selection of output response compression algorithm, test sequence generation and circuit implementation techniques are treated as a whole. This gave birth to a number of new digital testing techniques. Let us discuss some of them.

4.2. TRANSITION COUNT TESTING

The technique for testing digital circuits based on transition counting [57] was one of the first techniques to use integral estimates of output responses for testing and diagnosing digital circuits. It is based on the use of relationship (4.1), which allows one to produce the values of checksums (keywords) on the nodes of the circuit. As an example of transition count testing, consider the digital circuit of Fig. 4.1. The input lines to the circuit are applied with a test sequence $\{X_k\}$, $k = 1$ to l (Table 4.1) [57]. Each input pattern X_k, $k = 1$ to 5, is associated with output values generated at the nodes f_1, f_2, f_3 and f_4 of the circuit. If the circuit is fault-free, the checksums for all its nodes will match the reference ones (Table 4.1).

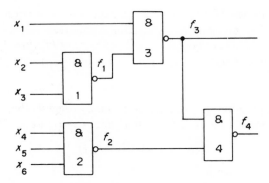

Fig. 4.1. A combinational network

On occurrence of a fault $x_1 \equiv 1$, the values of responses at the circuit nodes and their associated checksums will appear as follows: $f_1^* = 10100$, $f_2^* = 10111$, $f_3^* = 01011$, $f_4^* = 11100$, $R_{11}^* = 3$, $R_{12}^* = 2$, $R_{13}^* = 3$, $R_{14}^* = 1$. A change of pattern f_3 results in $R_{13}^* = 3$ and respectively $R_{13}^* \neq R_{13}$, i.e. the actual value of checksum R_{13}^* does not match the reference value R_{13}. This implies that a fault has occurred somewhere in the circuit under test, which is detectable at the output f_3 by transition count testing. At the same time, irrespective of $f_4 \neq f_4^*$, the equality $R_{14} = R_{14}^*$ shows that the fault is undetectable at output f_4.

An advantage of transition count testing is that the amount of data to be analysed at test application is significantly reduced. Thus, the number of bits required to represent R_1 or R_1^* does not exceed into ($\log_2 l$). Besides, the technique discussed is relatively simple in implementation of equipment, consisting of a transition detector, a counter and, as a rule, a count or pseudorandom sequence generator that is required for conducting the test experiment.

The main drawback of transition count testing is that fault coverage is reduced as compared with a classical approach based on matching the actual and reference responses of the circuit (Fig. 4.1). For the circuit of interest, the fault $x_1 \equiv 1$ cannot, be detected at output node f_4 by the present technique. However, by analysing whether $f_4 = 11000$ matches the actual value $f_4^* = 11100$ we can state that the above fault does exist. At the same time, when the sequence of test patterns X_k is changed by substituting X_4 for X_1 and X_1 for X_4, we obtain $f_4 = 01010$, $f_4^* = 01110$ and, hence, $R_{14} = 4$, $R_{14}^* = 2$, i.e. the fault is detectable by transition count testing. Hence, there are ways of obtaining reasonable test validity by the transition count technique.

Table 4.1.

X_k	x_1	x_2	x_3	x_4	x_5	x_6	f_1	f_2	f_3	f_4
X_1	1	0	1	0	0	0	1	1	0	1
X_2	1	1	1	1	1	1	0	0	1	1
X_3	0	1	0	1	1	0	1	1	1	0
X_4	1	1	1	1	0	1	0	1	1	0
X_5	0	1	1	0	1	1	0	1	1	0
R_1	3	1	2	2	3	3	3	2	1	1

Prior to discussing some of them, let us examine testability of an n-input NAND gate. All possible faults of the gate can be detected by the following test sequence [57]:

$$X_0 = 111 \cdots 1$$
$$X_1 = 011 \cdots 1$$
$$X_2 = 101 \cdots 1 \tag{4.5}$$
$$\vdots$$
$$X_n = 111 \cdots 0.$$

The reference response of the n-input NAND to a sequence of input sets $X_0, X_1, \ldots, X_n$ will appear as $f = 011 \cdots 1$. The difference between the actual response of the n-input NAND gate and the reference response implies that there exists a single or multiple stuck-at fault in the gate. Therefore, any test sequence, including the one used for implementation of the transition count testing, must contain all $(n + 1)$ input patterns (4.5).

As [57] reports, for $n = 1$ and $n = 2$, the sequence $X_0 X_1$ and $X_1 X_0 X_2$ are minimal test patterns that permit detection of all possible faults of an n-input NAND gate. However, even for $n > 2$, a minimal test will be the one consisting of a sequence of input patterns $X_0 X_1 X_2 X_3 \cdots X_n X_1$, whose number is $n + 2$. The reference value is $R_1 = 1$. Let us consider the possibility of detecting all possible faults by using the sequence $X_0 X_1 X_2 X_3 \cdots X_n X_1$. Depending on the type of fault, the following may occur.

(i) $a \equiv 1$ fault at the gate output is defined by the value $R_1^* = 0 \neq R_1$; hence the fault is detectable by comparing the actual checksum R_1^* with the reference checksum R_1.

(ii) $a \equiv 0$ fault appearing at the gate output causes the value $R_1^* = 0$ to be produced; hence it is also detectable.

(iii) a j-multiple fault $\equiv 1$ at the input nodes of an n-input NAND gate, where $1 \leqslant j \leqslant n - 1$, can cause the following values R_1^* to be produced. If the primary input line to the gate under test belongs to the set of inputs with the $\equiv 1$ fault, the output response of the circuit will take the form $f = 00d_2 d_3 \cdots d_n 0$ where $d_i \in \{0, 1\}$, $i = 2$ to n. Hence $R_1^* \geqslant 2$ and $R_1 \neq R_1^*$ respectively. If the primary input line does not belong to the set of fault inputs, the output response is represented by the sequence $f = 01d_2 d_3 \cdots d_n 1$ where at least one d_i is zero. In this case $R_1^* \geqslant 3$. As is evident, all possible stuck-at faults in an n-input NAND gate (with $n > 2$) can be detected by a sequence of $(n + 2)$ test patterns including all patterns from the set (4.5).

A major problem at implementing the transition count technique is to find the test sequences that can initiate generation of output responses that differ from the reference checksums of the circuit for any circuit modification. It is recommended then to use the values obtained by testing the minimal of all possible output responses of a digital circuit as reference checksums. In the general case, the number C of binary sequences of length l that have the identical value of checksum $R_1 = g$, where $0 \leqslant g \leqslant l - 1$, is expressed as

$$C = 2 \frac{(l - 1)!}{(l - 1 - g)! \, g!}. \tag{4.6}$$

By minimizing the value of C, we can improve the validity of testing and diagnosing of a digital circuit by transition count testing. In this case, the minimal number of possible faults will have checksums identical to the reference checksum. To reduce C, it must be made either minimally or maximally possible. Thus for $g = 0$, C assumes the value 2. This implies that only one output sequence of a digital circuit other than a response to the test sequence has a checksum identical to the reference one.

The above analysis allows one to lay down the design rule for digital circuits to be tested by the transition count technique [57].

To maximize the fault coverage of a transition count test, test sequences should be constructed so that the checksum R_1 (4.1) of their output responses is either as large or as small as possible.

For a digital combinational network, the order in which test patterns are applied is insignificant. Thus by rearranging the sequence of test patterns R_1 can be varied.

A test sequence generation technique for two-level digital circuits as well as fault location algorithms are discussed in [57].

A further development of transition count testing is the use of estimates R_3 and R_4 [66] to achieve better fault coverage for stuck-at faults in a circuit.

Although the technique of interest is simple in implementation, it has not found wide application. The so-called syndrome testing technique is more popular.

4.3. SYNDROME TESTING

The basic concepts of syndrome testing are in many ways similar to those of transition count testing discussed above. The distinctive property of syndrome testing (ones count testing) is in the fact that the syndrome (checksum) of an n-variable Boolean function is defined as

$$S = R_5/2^n \qquad (4.7)$$

where R_5 is estimated by (4.2) for $l = 2^n$ and equals the number of 1s for the function in accordance with the truth table. The definition of a syndrome uniquely implies that a count sequence generator should be used for generation of all possible binary combinations of n input variables on testing the circuit that implements the specified function [59].

By analysing expression (4.7) for computing syndrome S, we notice that $0 \leqslant S \leqslant 1$, and for the simplest logic circuits S assumes the values of Table 4.2.

In practice, it is useful to find the input–output syndrome relations of a logic circuit to estimate the values of syndromes at all nodes of the circuit. As an example, consider the relation between syndromes of a two-input OR gate.

Table 4.2.

	n-input gates					
Syndrome	AND	NAND	OR	NOR	EOR	NOT-EOR
S	2^{-n}	$1-2^{-n}$	$1-2^{-n}$	2^{-n}	2^{-1}	2^{-1}

The outputs of two functionally independent circuits with input n_1 and n_2, R_{51} and R_{52} values of 1s for functions implemented by them and syndromes S_1 and S_2 are connected to the inputs of an OR gate. To find the value of syndrome S_3, let us find the number R_{53} of 1s for the function implemented on the OR gate output. When computing R_{53}, note that the function describing the behaviour of the OR gate depends on $(n_1 + n_2)$ input variables. Then we have

$$R_{53} = R_{51}2^{n_2} + R_{52}2^{n_1} - R_{51}R_{52}.$$

Hence, subject to (4.7)

$$S_3 = \frac{R_{53}}{2^{n_1+n_2}} = \frac{R_{51}}{2^{n_1}} + \frac{R_{52}}{2^{n_2}} - \frac{R_{51}}{2^{n_1}}\frac{R_{52}}{2^{n_2}} = S_1 + S_2 - S_1S_2. \tag{4.8}$$

We can similarly demonstrate that for a two-input AND gate $S_3 = S_1S_2$, for a two-input EOR gate $S_3 = S_1 + S_2 - 2S_1S_2$ and for a NOT gate $S_3 = 1 - S_1$ [58].

The above relations associating the values of input and output syndromes allow one to estimate the syndromes of all nodes in a tree network.

As an example, consider the digital circuit of Fig. 4.2. Using the data of Table 4.2, we obtain $S_1 = 2^{-1}$, $S_2 = 1-2^{-3}$ and $S_3 = 2^{-2}$. According to equation (4.8) and the identical relationship for a NOT gate, we find that

$$S_4 = 1 - (S_2 + S_3 - S_2S_3)$$
$$= 1 - (1 - 2^{-3} + 2^{-2} - 2^{-2} + 2^{-5}) = 3/32.$$

The syndrome value is $S_5 = S_1S_4 = 3/64$.

More complex is the procedure of calculating syndromes for the general case of a digital circuit with fan-in branchings as shown in [58]. Prior to solving the task, let us define the properties of syndrome $S(F)$ of a switching function $F(x_1, \cdots, x_n)$.

(i) According to equation (4.7), $0 \leqslant S(F) \leqslant 1$ with $S(F) = 0$ if $F(x_1,...,x_n) = 0$ and $S(F) = 1$ if $F(x_1,...,x_n) = 1$.

(ii) $S(F) = 1 - S(\bar{F})$ as implied by equation (4.7)

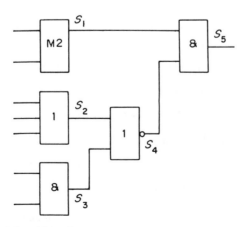

Fig. 4.2. A tree circuit for which reference values of syndromes S_1, S_2, S_3, S_4 and S_5 are defined

To calculate the syndromes of complex switching functions, let us define the relation for the syndrome of ORed functions $F(x_1, \cdots, x_n)$ and $G(x_1, ..., x_n)$ for which p variables are shared with $p \leqslant \min(n_1, n_2)$. Let us prove the following theorem [59].

Theorem 4.1. The syndrome of ORed switching functions F and G is defined by

$$S(F + G) = S(F) + S(G) - S(FG) \qquad (4.9)$$

where ANDing and ORing are performed on Boolean functions and arithmetic addition and subtraction are performed on syndromes.

Proof. Suppose the digital circuit realizes the Boolean function $F + G$ where input variables are $x_1, ..., x_p, ..., x_{n_1}, ..., n_{n_2}$, with $n_2 \geqslant n_1$. The total number of input lines to the circuit is $n_1 + n_2 - p$, where p lines may be shared by both circuits realizing functions F and G. According to the truth table, the number of 1s for function $F + G$ can be determined by [58]

$$R_s(F + G) = R_s(F)2^{n_2 - p} + R_s(G)2^{n_1 - p} - R_s(FG).$$

Considering that the number of variables is $n_1 + n_2 - p$, divide the right-hand and left-hand sides of the obtained equation by $2^{n_1 + n_2 - p}$ to give

$$\frac{R_s(F + G)}{2^{n_1 + n_2 - p}} = \frac{R_s(F)}{2^{n_1}} + \frac{R_s(G)}{2^{n_2}} - \frac{R_s(FG)}{2^{n_1 + n_2 - p}}.$$

Hence, subject to (4.7), we shall finally obtain the expressions (4.9), as we wished to prove. ■

The obtained expression for calculating the syndrome of ORed Boolean functions (4.9) is more general than (4.8), which can be treated as its particular case with $p = 0$.

Theorem 4.2. The syndrome of the logic product of switching functions F and G is calculated by

$$S(FG) = S(F) - S(F\bar{G}). \qquad (4.10)$$

Proof. Define the syndrome of function F, which is represented in the form $F = FG + F\bar{G}$. Then according to (4.9), we obtain

$$S(FG + F\bar{G}) = S(FG) + S(F\bar{G}) - S(FGF\bar{G}) = S(FG) + S(F\bar{G}) - S(0).$$

Considering that $S(0)$ is zero by property (i), we finally obtain

$$S(F) = S(FG) + S(F\bar{G})$$

whence by conversion we obtain the relation (4.10) as was to be shown. ■

From this property of the product syndrome the following relation is obtained

$$S(F\bar{G}) = S(F) - S(FG). \qquad (4.11)$$

Theorem 4.3. The modulo-2 sum syndrome of logic functions F and G is calculated by

$$S(F \oplus G) = S(F) + S(G) - 2S(FG). \qquad (4.12)$$

Proof. Represent $F \oplus G$ in the form $F\bar{G} + \bar{F}G$. The syndrome of $F\bar{G} + \bar{F}G$, according to (4.9), will be determined as

$$S(F\bar{G} + \bar{F}G) = S(F\bar{G}) + S(\bar{F}G) - S(F\bar{G}\bar{F}G) = S(F\bar{G}) + S(\bar{F}G).$$

Taking account of the corollary to Theorem 4.2, we finally obtain

$$S(F \oplus G) = S(F) - S(FG) + S(G) - S(FG) = S(F) + S(G) - 2S(FG)$$

as was to be shown. ∎

When the functions F and G are independent of shared variables, i.e. $p = 0$, the expression (4.12) takes the form

$$S(F \oplus G) = S(F) + S(G) - 2S(F)S(G).$$

As an example of syndrome calculation, let us consider the digital circuit of Fig. 4.3, which realizes the switching function $F = x_1\bar{x}_2\bar{x}_3 + \bar{x}_1x_2x_3$. The syndrome of the function is defined by

$$\begin{aligned}
S(F) &= S(x_1\bar{x}_2\bar{x}_3 + \bar{x}_1x_2x_3) \\
&= S(x_1\bar{x}_2\bar{x}_3) + S(\bar{x}_1x_2x_3) - S(x_1\bar{x}_2\bar{x}_3\bar{x}_1x_2x_3) \\
&= 1/2^3 + 1/2^3 - 0 \\
&= 1/4.
\end{aligned}$$

At the same time, the syndrome of the function that describes the behaviour of the circuit of Fig. 4.3 having fault $x_1 \equiv 1$ is $1/4$. Actually, with $x_1 \equiv 1$, $F = \bar{x}_2\bar{x}_3$ and $S(F) = S(\bar{x}_2\bar{x}_3) = 1/4$. This implies that the fault $\equiv 1$ at the input x_1 of the circuit in Fig. 4.3 is undetectable by comparing the reference syndrome with the actual syndrome.

Let us state the conditions for the digital circuit defined by a Boolean function F to be syndrome-testable, i.e. all its faults are detectable by comparing the actual and reference syndromes. Consider two versions of F dependence on x_i [59]. In the first case, function $F = F(x_1, \ldots, x_n)$ can be represented by

$$F = G_1x_i + G_2 \qquad i \in \{1, 2, 3, \ldots, n\} \tag{4.13}$$

where $G_1 \neq 0$ and $G_2 \neq 1$ as well as $\delta G_1/\delta x_i = \delta G_2/\delta x_i = 0$. The syndrome is expressed

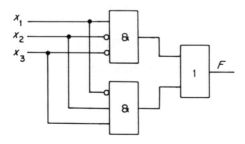

Fig. 4.3. A digital circuit that realizes the switching function $F = x_1\bar{x}_2\bar{x}_3 + \bar{x}_1x_2x_3$

by

$$S(F) = S(G_1 x_i + G_2)$$
$$= S(G_1 x_i) + S(G_2) - S(G_1 G_2 x_i)$$
$$= S(G_2) + S(G_1 \bar{G}_2 x_i)$$
$$= S(G_2) + S(G_1 \bar{G}_2)/2.$$

Any fault $x_i \equiv 0$ causes the syndrome of function F_0^*, which describes the behaviour of a faulty modification of the circuit to be $S(F_0^*) = S(G_2)$. This fault is undetectable with $S(F) = S(F_0^*)$, which is only possible with $S(G_1 \bar{G}_2) = 0$. However, by property (i), the relation $S(G_1 \bar{G}_2) = 0$ holds only for the function $G_1 \bar{G}_2 = 0$, which contradicts the adopted conditions $G_1 \neq 0$ and $G_2 \neq 1$ defining the non-redundant digital circuit against variable x_i.

For a fault $x_i \equiv 1$, the syndrome of F_1^* is defined by

$$S(F_1^*) = S(G_1 + G_2) = S(G_1) + S(G_2) - S(G_1 G_2) = S(G_2) + S(G_1 \bar{G}_2).$$

Hence $S(F) \neq S(F_1^*)$ for $x_i \equiv 1$. Therefore any stuck-at fault at the input line x_i to the circuit described by Boolean function $F = G_1 x_i + G_2$ (4.13) is syndrome-testable.

The result obtained for (4.13) can be easily extended to a more general case which can be stated as the following theorem [59].

Theorem 4.4. A two-level non-redundant digital circuit described by a Boolean function $F(x_1, ..., x_n)$ and dependent only on a variable $\bar{x}_1$ or on its negative x_i, $i = 1$ to n, is syndrome-testable at the input line that is applied with the variable or its negation.

In the general case, the Boolean function $F(x_1, ..., x_n)$ does not depend on x_i or $\bar{x}_i$. Most often it can assume the form

$$F = G_1 x_i + G_2 \bar{x}_i + G_3 \qquad i \in \{1, 2, 3, ..., n\} \tag{4.14}$$

where $G_1 \neq 0$, $G_2 \neq 0$, $G_3 \neq 1$ and $\delta G_1/\delta x_i = \delta G_2/\delta x_i = \delta G_3/\delta x_i = 0$. The syndrome of the function is defined by

$$S(F) = [S(G_1 \bar{G}_3) + S(G_2 \bar{G}_3)]/2 + S(G_3). \tag{4.15}$$

For the fault $x_i \equiv 1$, the initial function F is converted to $F_1^* = G_1 + G_3$ whose syndrome is calculated by

$$S(F_1^*) = S(G_1 \bar{G}_3) + S(G_3) \tag{4.16}$$

and for the fault $x_i \equiv 0$ it is respectively converted to $F_0^* = G_2 + G_3$ and its syndrome is calculated by

$$S(F_0^*) = S(G_2 \bar{G}_3) + S(\bar{G}_3). \tag{4.17}$$

The analysis of the obtained values $S(F)$, $S(F_0^*)$ and $S(F_1^*)$ shows that resulting syndromes $S(F_0^*)$ and $S(F_1^*)$ can be identical to the reference $S(F)$ if and only if the condition

$$S(G_1 \bar{G}_3) = S(G_2 \bar{G}_3) \tag{4.18}$$

is met. The above-mentioned faults are syndrome-untestable only when equation

(4.18) is satisfied. At the same time, the inequality

$$S(G_1\bar{G}_3) \neq S(G_2\bar{G}_3) \tag{4.19}$$

can be the condition for the fault $x_i \equiv 1$ or $x_i \equiv 0$ to be testable by comparing the actual syndromes with the reference ones. Let us state the property as a theorem.

Theorem 4.5. A two-level non-redundant digital circuit described by a Boolean function $F = G_1 x_i + G_2 \bar{x}_i + G_3$, $i \in \{1, 2, 3,..., n\}$, where $G_1 \neq 0$, $G_2 \neq 0$ and $G_3 \neq 1$, and $\delta G_1/\delta x_i = \delta G_2/\delta x_i = \delta G_3/\delta x_i = 0$, is syndrome-testable at the input line where variable x_i is applied if and only if the inequality $S(G_1\bar{G}_3) \neq S(G_2\bar{G}_3)$ is satisfied.

Theorem 4.5 is a more general statement than theorem 4.4 since the testability condition $S(G_1\bar{G}_3) \neq S(G_2\bar{G}_3)$ is rigorously met with $G_1 = 0$ and $G_2 \neq 0$ or $G_1 \neq 0$ and $G_2 = 0$, thereby corresponding to the formulation of the latter.

Theorem 4.5 allows one to state whether a digital circuit is testable by comparing the syndromes as well as to lay down the design rules for syndrome-testable digital circuits. Let us consider some of them.

4.4. DESIGN OF SYNDROME-TESTABLE DIGITAL CIRCUITS

Among the first approaches to designing syndrome-testable digital circuits was the one based on inserting extra inputs to an AND gate in a two-level non-redundant circuit. The extra inputs are used in the test and diagnostic mode. In the normal mode, these inputs are usually fed with logic 1s [58, 59].

The basic idea behind insertion of extra inputs is to make the circuit testable at the input where equation (4.18) is satisfied.

Syndrome-testable circuit design algorithms based on inserting extra inputs, which have been detailed in [58, 59], use the property of syndromes formulated as follows.

Theorem 4.6. If a Boolean function $F = F(x_1,..., x_n) = G_1 x_i + G_2 \bar{x}_i + G_3$, $i \in \{1, 2, 3,..., n\}$, where $G_1 \neq 0$, $G_2 \neq 0$, $G_3 \neq 1$, $\delta G_1/\delta x_i = \delta G_2/\delta x_i = \delta G_3/\delta x_i = 0$, is syndrome-untestable in x_i then the Boolean function $F_1 = F_1(x_i,..., x_n, c_1) = c_1 G_1 x_i + G_2 \bar{x}_i + G_3$ is syndrome-testable in the given variable.

Proof. To define testability of function $F_1 = c_1 G_1 x_i + G_2 \bar{x}_i + G_3$, we must determine whether condition (4.19), which has the form

$$S(c_1 G_1 \bar{G}_3) \neq S(G_2 \bar{G}_3) \tag{4.20}$$

for the given function, is satisfied. By applying relation (4.10) to the left-hand side of the given inequaliity, we obtain

$$S(G_1\bar{G}_3) - S(G_1\bar{G}_3\bar{c}_1) \neq S(G_2\bar{G}_3).$$

Subject too equation (4.18), which holds true for the function $F = G_1 x_i + G_2 \bar{x}_i + G_3$, we finally obtain

$$- S(G_1\bar{G}_3\bar{c}_1) \neq 0.$$

This implies that the testability of $F_1 = F_1(x_i,...,x_n,c_1)$ in x_i is rigorously satisfied, as was to be shown.   ∎

Note that $F_1(x_1,...,x_n,c_1) = F(x_1,...,x_n)$ for $c_1 = 1$; therefore input c_1 to the circuit that implements the function F is used only in the test and diagnostic mode. In the normal mode, $c_1 = 1$.

An example of the procedure of designing a syndrome-testable digital circuit is the Boolean function $F = x_1\bar{x}_2\bar{x}_3 + \bar{x}_1 x_2 x_3$, whose implementation is syndrome-untestable in all variables (Fig. 4.3). To make the circuit that implements F testable, let us introduce an extra variable c_1 to be fed to either input of its first-level AND gate. The modified function $F_1 = F_1(x_1,...,x_n,c_1)$ takes the form $F_1 = c_1 x_1\bar{x}_2\bar{x}_3 + \bar{x}_1 x_2 x_3$ whose testability conditions with respect to variables x_1, x_2 and x_3 may be the following:

$$S(c_1\bar{x}_2\bar{x}_3) \neq S(x_2 x_3)$$
$$S(x_1\bar{x}_3) \neq S(c_1\bar{x}_1 x_3)$$
$$S(x_1\bar{x}_2) \neq S(c_1\bar{x}_1 x_2).$$

The above results are used for an arbitrary digital circuit rather than the classical version of its two-level implementation [58, 59]. However, this case requires both extra inputs and outputs to be inserted [59]. Thus in the example of [59], the arithmetic device consisting of more than 100 gates requires a single extra input and two extra outputs for its implementation.

A major disadvantage of syndrome-testable digital circuit design is the insertion of extra external nodes, since the problem of external nodes is very urgent for LSI/VLSI chips. Besides, extra inputs make the testing procedure more time-consuming.

Another approach to testing syndrome-untestable digital circuits is that based on a modification of the algorithm for generation of input test patterns[67]. Consider the approach for the general case of a digital circuit defined by a Boolean function $F = F(x_1,...,x_n)$ with respect to any node $g = g(x_1,...,x_n)$. For the purpose, let us represent the function F as

$$F = G_1 g + G_2 \bar{g} + G_3$$

where $G_1 \neq 0$, $G_2 \neq 0$, $G_3 \neq 1$, $\delta G_1/\delta g = \delta G_2/\delta g = \delta G_3/\delta g = 0$. The syndrome-untestable condition for $g \equiv 0$ is defined by

$$S(G_1\bar{G}_3 g) = S(G_2\bar{G}_3 g)$$

and for $g \equiv 1$ by [67]

$$S(G_1\overline{G_3 g}) = S(G_2\overline{G_3 g})$$

whereas syndrome-testable conditions for the same faults on node g are defined respectively as

$$S(G_1\bar{G}_3 g) \neq S(G_2\bar{G}_3 g) \qquad S(G_1\overline{G_3 g}) \neq S(G_2\overline{G_3 g}). \qquad (4.21)$$

They are rigorously satisfied for the following relations of values for functions G_1 and G_2: $G_1 \neq 0$, $G_2 = 0$; $G_1 \neq 1$, $G_2 = 1$; $G_1 = 0$, $G_2 \neq 0$; and $G_1 = 1$, $G_2 \neq 1$ [67]. A solution that gives the required connection between functions G_1 and G_2 is to define. The values of variables in the assigned sets of circuit inputs whose existence has been proved in [67]. Therefore, when conditions $G_1 \neq 0$ and $G_2 = 0$ are met, the source function can be represented as $F = G_1 g + G_3$. For the above conditions of Theorem 4.4, all faults on node g are syndrome-testable, which also holds true for its non-trivial implementation as a multi-level circuit [59].

A disadvantage of the technique for syndrome-untestable digital circuits is irregularity of the input test pattern generation algorithm, which makes the diagnostic procedure more complex. Thus, for LSI and VLSI, the design of self-testing circuits is impracticable. More applicable is a circuit modification that ensures testability by comparing the reference and actual syndromes. The circuit modification that allows one to achieve syndrome testability for VLSI chips has been discussed in [68].

Syndrome testing can be further developed by combining it with the techniques based on scanning the states of memory elements (see Sect. 3.2) for implementing self-testable VLSI chips. A more promising approach is the use of syndrome testing for LSSD VLSI chips [60]. In this case, the original VLSI chip is partitioned into several subcircuits with a small number of inputs. Their responses are tested by the so-called weighted syndrome due to which the test time is substantially reduced for VLSI chips.

4.5. DESIGN AND VERIFICATION OF CIRCUITS WITH ODD NUMBER OF 1s IN OUTPUT RESPONSES

A design technique for digital circuits has been proposed for use in conjunction with the LSSD technique and is based on the following assumptions.

(i) A digital circuit described by a Boolean function $F = F(x_1, ..., x_n)$ is a single-input circuit.

(ii) At testing, the circuit inputs are applied with all possible binary combinations of variables $x_1, ..., x_n$, with each combination being applied only once.

(iii) At testing, an odd number of 1s is generated on the circuit output if the circuit is free of stuck-at faults [69].

The latter assumption implies that an odd number of 1s should be provided at the output of the reference digital circuit. For the majority of circuits the requirement is satisfied automatically; however, in some cases the circuit needs some modification. A simple modification of the digital circuit is the use of an extra n-input AND gate [70].

The concept estimation of the circuit output response is a value indicating whether the number of 1s is even or odd. It can be obtained at a memory element that performs modulo-2 summing on data stored by the element and applied to its input.

The essence of the discussed technique for design and verification of digital circuits with odd number of 1s in their output responses is disclosed by the following theorem [69].

Theorem 4.7. During the test, all possible combinations of $\equiv 0$ and $\equiv 1$ faults at the circuit inputs and output as well as many other faults are detectable in a single-input circuit that meets the adopted assumptions.

Proof. When $a \equiv 0$ or $\equiv 1$ fault occurs at the output of the circuit described by a

Boolean function $F = F(x_1, \ldots, x_n)$ its faulty modifications will be described respectively by $F_1^* = 1$ and $F_0^* = 0$. Then 2^n 1s and 2^n 0s are generated respectively on the circuit output, with the number of elements being even in both cases. Therefore, the specified faults are testable.

When $a \equiv 0$ or $\equiv 1$ appears on input x_1 the faulty circuit modification is described by a Boolean function $F^* = F^*(x_1, \ldots, x_{i-1}, x_{i+1}, \ldots, x_n)$, which depends on 2^{n-1} variables. However, its output values will be repeated twice on applying 2^n input test patterns of variables $x_1, \ldots, x_{i-1}, x_i, x_{i+1}, \ldots, x_n$ and the number of 1s for $F^*(x_1, \ldots, x_{i-1}, x_{i+1}, \ldots, x_n)$ will become even.

Thus the discussed faults are testable by testing the circuit output response for parity. When a two-, three- or n-fold fault occurs at an input, the number of 1s in the output response of the faulty circuit modification must be multiplied respectively by 4, 8 or 2^μ, where μ is the fault multiplicity. The test result will always be even, thereby indicating to the existence of a fault.

We can similarly show that many other faults that produce an output string with an even number of 1s in the circuit are also testable, as was to be shown. ∎

Let us express the efficiency of the discussed technique as a percentage of probability p for an arbitrary fault in a digital circuit to be testable [69].

Let the circuit that implements the Boolean function $F = F(x_1, \ldots, x_n)$ have $k > n$ nodes where stuck-at-0 and stuck-at-1 faults are possible. Then the number of possible faults for the circuit is $3^k - 1$. When a stuck-at-0 or stuck-at-1 fault occurs at the primary input of the circuit, all the faults will be testable irrespective of the state of the other $(k - 1)$ nodes. Taking into account that input x_1 can be in two faulty conditions whereas the remaining $(k - 1)$ nodes can be in one of three ($\equiv 0$, $\equiv 1$, good) conditions, the total number of testable faults will be $2 \times 3^{k-1}$. We can similarly show that for a stuck-at fault at the jth input of the circuit the number of testable faults approaches $2 \times 3^{k-j}$ due to the absence of faults at its $(j - 1)$ primary inputs. The total number of testable faults can be found by

$$2 \sum_{j=1}^{n} 3^{k-j} + 2 \times 3^{k-n-1} \tag{4.22}$$

where the latter term is obtained due to a fault occurrence at the circuit output. The probability of fault testability is calculated as the ratio of the detected faults to the total number of probable faults:

$$p = \frac{2\sum_{j=1}^{n} 3^{k-1} + 2 \times 3^{k-n-1}}{3^k - 1} = \frac{3^k - 3^{k-n-1}}{3^k - 1}.$$

An estimation of p can be the relation

$$p > 1 - \frac{1}{3^{n+1}}$$

where p is a function of circuit inputs, with $p = 0.995\,8848$ even for $n = 4$ and $p = 0.999\,9999$ for $n = 15$.

An example of the discussed testing technique can be the circuit of Fig. 4.3, which is to be modified to make the number of 1s on its output odd. An additional n-input AND gate, $n = 4$, to be used only in the test mode, ensures that by applying an

Table 4.3.

x_1	x_2	x_3	$F = F(x_1, \ldots, x_n)$	$F^* = F(x_1 \equiv 1)$	$F^* = F(x_1 \equiv 0)$
0	0	0	1	1	1
0	0	1	0	0	0
0	1	0	0	0	0
0	1	1	1	0	1
1	0	0	1	1	1
1	0	1	0	0	0
1	1	0	0	0	0
1	1	1	0	0	1

enabling signal $c_1 = 1$ we obtain a circuit with an odd number of 1s generated at its output. The values of output symbols obtained as a result of applying all possible combinations to the circuit inputs are given in Table 4.3 for fault-free and faulty conditions.

The analysis of output circuit responses $x_1 \equiv 1$ and $x_1 \equiv 0$ shows that the specified faults are testable by parity checking on 1s in its output sequences. It should be noted, however, that stuck-at-0 faults occurring at the outputs of any pair of AND gates are undetectable. The discussed method can be further enhanced by using it in conjunction with syndrome testing.

5

RANDOM TESTING

5.1. CONCEPT OF RANDOM TESTING

A major feature of random testing is the use of random sequences of independent bits applied to the inputs of a circuit under test. Here a variable $x_i \in \{0, 1\}$, $i = 1$ to n, applied to the ith input is described by probability $P(x_i = 1)$ of its value to be 1.

The classical random testing network consists of a random bit sequence generator, the circuit under test and a reference circuit as well as an output response comparator. The bit sequence generator is used to produce independent random numbers by the specified probability distribution $P(x_i = 1)$. Binary patterns produced at the generator output are applied simultaneously to the inputs of the circuit under test and the reference circuit. Output responses of these circuits are compared in the output response comparator. When the output responses from both circuits are identical, the circuit under test is considered to be fault-free; otherwise it is assumed to be faulty and the problem of diagnosing it arises.

Publications on random testing [63, 71] deal with finding the reference values for the probability of 1s to appear in the intermediate and output nodes of the circuit. By the proposed techniques, the reference probability values that have been obtained analytically were compared with the actual values. Thus, the random testing procedure consists of obtaining actual probability values and their comparison with reference values. The mentioned techniques have not found wide application due to the necessity of calculating reference probability values for a 1 to appear in the intermediate and output nodes of the circuit.

Consider the relationship between output probabilities for simple logic devices and probabilities $P(x_i = 1)$, $i = 1$ to n, for their input variables $x_1, x_2, \ldots, x_n$. Considering $x_i \in \{0, 1\}$, we obtain

$$P(x_i = 1) = 1 - P(x_i = 0). \tag{5.1}$$

Then examine a NOT gate whose input variable x_1 uniquely determines the output value y, where $y = \bar{x}_1$. Since the probability $P(y = 1) = P(\bar{x}_1 = 1)$, then, subject to (5.1), we obtain [72]

$$P(y = 1) = P(x_1 = 0) = 1 - P(x_1 = 1). \tag{5.2}$$

To examine the probability relation $P(y = 1)$ of the AND gate, apply sequences of random binary numbers that are distinct magnitudes to its inputs. The output value of the AND gate is 1 if and only if all input variables are 1. The probability of the event is determined as [72]

$$P(y = 1) = P(x_1 = 1, x_2 = 1, \ldots, x_n = 1)$$

where $y = x_1 x_2 \cdots x_n$. Considering the events $x_1 = 1$, $x_2 = 1, \ldots, x_n = 1$ to be independent, the latter relationship can be rearranged as

$$P(y = 1) = \sum_{i=1}^{n} P(x_i = 1). \tag{5.3}$$

For a two-input AND gate, $P(y = 1) = P(x_1 = 1)P(x_2 = 1)$.

Next consider a multi-input OR gate whose output value can be obtained by the expression

$$y = x_1 + x_2 + \cdots + x_n = \overline{\overline{x_1 + x_2 + \cdots + x_n}} = \overline{\bar{x}_1 \bar{x}_2 \cdots \bar{x}_n}.$$

Applying (5.2) to $y = \overline{\bar{x}_1 \bar{x}_2 \cdots \bar{x}_n}$, we obtain

$$P(y = 1) = P(\overline{\bar{x}_1 \bar{x}_2 \cdots \bar{x}_n} = 1) = 1 - P(\bar{x}_1 \bar{x}_2 \cdots \bar{x}_n = 1).$$

Taking into account that the events $\bar{x}_1 = 1, \bar{x}_2 = 1, \ldots, \bar{x}_n = 1$ are independent and applying (5.3) we can finally obtain

$$P(y = 1) = 1 - \sum_{i=1}^{n} P(\bar{x}_i = 1)$$

$$= 1 - \sum_{i=1}^{n} [1 - P(x_i = 1)]. \tag{5.4}$$

From equation (5.4)

$$P(y = 1) = P(x_1 = 1) + P(x_2 = 1) - P(x_1 = 1)P(x_2 = 1)$$

for the two-point OR gate.

When the events $x_i = 1$ and $x_j = 1$ are not simultaneous, with $i \neq j$, relation (5.4) is rearranged as

$$P(y = 1) = \sum_{i=1}^{n} P(x_i = 1). \tag{5.5}$$

A more complete list of relationships between output probabilities of gates as probability functions of output variables and their correlation functions is given in [72]. In these relationships the expressions for AND and OR gates can assume quite different form from those of (5.3) and (5.4). Thus, for example, for a two-input AND gate with x_1 applied to both its inputs, the output probability $P(y = 1)$ is given by [73]

$$P(y = 1) = P(x_1 x_1 = 1) = P(x_1 = 1).$$

We can similarly show that

$$P(x_1 \bar{x}_1 = 1) = 0 \qquad P(x_1 + x_1 = 1) = P(x_1 = 1) \qquad P(x_1 + \bar{x}_1 = 1) = 1.$$

However, in the most simple case of a combinational tree, relations (5.2)–(5.4) will be sufficient for finding probability values for intermediate and output nodes of the digital circuit. As an example, consider the circuit in Fig. 5.1 whose inputs are fed with the sequences of independent random bits with probabilities $P(x_1 = 1) = P(x_2 = 1) = P(x_3 = 1) = P(x_4 = 1) = P(x_5 = 1) = p$. By applying relations (5.2)–(5.4), we obtain $P(f_1 = 1) = p^3$ and $P(f_2 = 1) = 1 - 2p + p^2$, respectively. For the output value

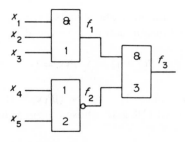

Fig. 5.1. A digital implementing the switching function $f_3 = x_1 x_2 x_3 \overline{(x_4 + x_5)}$

of the circuit, we finally obtain

$$P(f_3 = 1) = P(f_1 = 1)P(f_2 = 1) = p^3(1 - 2p + p^2) = p^3 - 2p^4 + p^5.$$

With $p = \frac{1}{2}$, the reference probability values take the form

$$P(f_1 = 1) = 0.125 \qquad P(f_2 = 1) = 0.25 \qquad P(f_3 = 1) = 0.031\,25.$$

When random testing is realized for the circuit under test, the actual probability values $P(f_1 = 1)$, $P(f_2 = 1)$ and $P(f_3 = 1)$ are compared with the reference values and then a decision on the circuit state is made. For a faulty condition, the actual probability values normally differ from their reference values. Thus, for example, for the fault $f_1 \equiv 1$, we obtain $P^*(f_1 = 1) = 1$, $P^*(f_2 = 1) = 0.25$, $P^*(f_3 = 1) = 0.25$, indicating that the circuit is faulty.

In the general case, estimation of reference probabilities is a complex analytical task, especially for digital circuits of arbitrary form. Here special techniques are required. Let us discuss some of them.

5.2. PROBABILISTIC ANALYSIS TECHNIQUES FOR DIGITAL CIRCUITS

Two algorithms discussed in [73] were among the first probabilistic analysis techniques for digital circuits. They make use of earlier relations (5.2)–(5.4) and are applied for combinational networks.

The first algorithm has the following properties [61].

(i) A Boolean function that describes any node in a combinational network can be represented in the sum-of-products form of input variables and their negations.

(ii) A Boolean function that has been expressed in a sum-of-products form can be represented as a canonical sum of products [73]:

$$\pi_i \pi_j = 0 \qquad i \neq j.$$

A straightforward algorithm for finding the probability of a 1 signal to appear in any node of the network consists of the following steps.

(i) Find the analytical expression for the specified node of the network as a canonical sum of products

$$F = \bigvee_{i=1}^{k} \pi_i.$$

(ii) Estimate $P(\pi_i = 1)$ for any product π_i by using relations (5.2) and (5.3).

(iii) Since the events $\pi_i = 1$ and $\pi_j = 1$, $i \neq j$, are not concurrent, estimate $P(F = 1)$ by equation (5.5).

As an example of the algorithm discussed, let us find $P(F = 1)$, where $F = (x_1 + x_2)x_3 + x_1 x_2$, for the network of Fig. 5.2 [73].

(i) The Boolean function $F = x_1 x_2 + x_1 x_3 + x_2 x_3$ is represented as a canonical sum of products

$$F = x_1 x_2 x_3 + \bar{x}_1 x_2 x_3 + x_1 \bar{x}_2 x_3 + x_1 x_2 \bar{x}_3.$$

(ii) Find sequentially the values $P(x_1 x_2 x_3 = 1)$, $P(\bar{x}_1 x_2 x_3 = 1)$, $P(x_1 \bar{x}_2 x_3 = 1)$ and $P(x_1 x_2 \bar{x}_3 = 1)$. Then

$$P(x_1 x_2 x_3 = 1) = P(x_1 = 1)P(x_2 = 1)P(x_3 = 1)$$
$$P(\bar{x}_1 x_2 x_3 = 1) = [1 - P(x_1 = 1)]P(x_2 = 1)P(x_3 = 1)$$
$$P(x_1 \bar{x}_2 x_3 = 1) = P(x_1 = 1)[1 - P(x_2 = 1)]P(x_3 = 1)$$
$$P(x_1 x_2 \bar{x}_3 = 1) = P(x_1 = 1)P(x_2 = 1)[1 - P(x_3 = 1)].$$

(iii) According to (5.5), we find

$$P(F = 1) = P(x_1 x_2 x_3 = 1) + P(\bar{x}_1 x_2 x_3 = 1) + P(x_1 \bar{x}_2 x_3 = 1) + P(x_1 x_2 \bar{x}_3 = 1)$$
$$= P(x_1 = 1)P(x_2 = 1) + P(x_1 = 1)P(x_3 = 1) + P(x_2 = 1)P(x_3 = 1)$$
$$- 2p(x_1 = 1)P(x_2 = 1)P(x_3 = 1).$$

For $P(x_1 = 1) = P(x_2 = 1) = P(x_3 = 1) = p$, we obtain

$$P(F = 1) = 3p^2 - 2p^3.$$

A major disadvantage of the technique discussed is the complexity of representing the analytical expression as a canonical sum of products. The second algorithm that allows one to obtain the probability representation of a digital circuit on the basis of its functional diagram is easier to use [73]. This algorithm relies on the probability

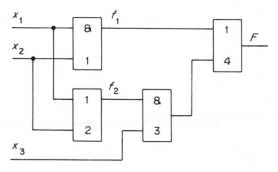

Fig. 5.2. A combinational circuit

property of dependent events. Let us consider the said property by using a five-input AND gate as an example. Suppose three of its primary inputs are applied with variable x_1 whose probability to be 1 is $P(x_1 = 1)$. The fourth and fifth primary inputs are applied with x_2 and x_3 whose probabilities are $P(x_2 = 1)$ and $P(x_3 = 1)$, respectively. According to (5.3), the probability $P(F = 1)$ of a 1 to appear at the output of the five-input AND gate is defined by the expression

$$P(F = 1) = P^3(x_1 = 1)P(x_2 = 1)P(x_3 = 1)$$

where $P(x_1 = 1)$ has degree 3 in the latter expression. On the other hand, from the AND gate behaviour we may deduce that the probability of a 1 to appear at its output is characterized by the probability concurrent events $x_1 = 1$, $x_2 = 1$ and $x_3 = 1$, which is calculated for independent variables x_1, x_2 and x_3 as a product of probabilities:

$$P(F = 1) = P(x_1 = 1)P(x_2 = 1)P(x_3 = 1)$$

where all probabilities have degree 1 only. Thus, for obtaining the correct probability estimate, which describes the behaviour of a combinational circuit with reconvergent fan-out (dependent events), we must reduce the degrees of probabilities to 1 in the probability expression obtained by using (5.2)–(5.4) [73].

The algorithm that makes use of the property consists of the following steps.

(i) All the nodes of the digital circuit under test are assigned different variables.

(ii) For each element of the circuit, the output probability is calculated as a function of input variable probabilities with the procedure being performed in succession from the elements to which the circuit inputs are connected to the element with the function implemented by the circuit produced on its output.

(iii) To obtain the correct probability expression for the circuit, all powers are reduced to 1.

As an example of the above algorithm, let us find $P(F = 1)$ for the circuit of Fig. 5.2. By using the relations (5.3) and (5.4), we obtain

$$P(f_1 = 1) = P(x_1 = 1)P(x_2 = 1)$$
$$P(f_2 = 1) = P(x_1 = 1) + P(x_2 = 1) - P(x_1 = 1)P(x_2 = 1)$$
$$P(f_3 = 1) = P(f_2 = 1)P(x_3 = 1) = P(x_1 = 1)P(x_3 = 1) + P(x_2 = 1)P(x_3 = 1)$$
$$- P(x_1 = 1)P(x_2 = 1)P(x_3 = 1)$$
$$P(F = 1) = P(f_1 = 1) + P(f_3 = 1) - P(f_1 = 1)P(f_3 = 1) = P(x_1 = 1)P(x_2 = 1)$$
$$+ P(x_1 = 1)P(x_3 = 1) + P(x_2 = 1)P(x_3 = 1) - P(x_1 = 1)P(x_2 = 1)P(x_3 = 1)$$
$$- P^2(x_1 = 1)P(x_2 = 1)P(x_3 = 1) - P(x_1 = 1)P^2(x_2 = 1)P(x_3 = 1)$$
$$+ P^2(x_1 = 1)P^2(x_2 = 1)P(x_3 = 1).$$

Reducing the powers to 1, we finally have

$$P(F = 1) = P(x_1 = 1)P(x_2 = 1) + P(x_1 = 1)P(x_3 = 1) + P(x_2 = 1)P(x_3 = 1)$$
$$- 2P(x_1 = 1)P(x_2 = 1)P(x_3 = 1).$$

The problem of finding the probabilistic expression for a digital circuit can be substantially simplified for a circuit employing logic gates of the same type. As has been demonstrated in [74, 75], for a tree circuit employing n-input NAND gates with the inputs of each gate connected to the outputs of the preceding level gates, the probability expression will have the form [75].

$$P_l(y=0) = P_{l-1}^n(y=1) = [1 - P_{l-1}(y=0)]^n \qquad (5.6)$$

where $P_l(y=0)$ and $P_l(y=1)$ are respectively the probabilities of a 0 and 1 to appear on the outputs of the lth-level NAND gates.

An original algorithm based on transforming the source digital circuit to a tree structure has been proposed in [76]. The idea behind the approach is to establish the bounds for the sought-for probability by simulating limiting probabilities in fan-out points. In this case, a fan-out is represented as independent inputs with the probability of a 1 to occur being in the range 0 to 1. Thus for the example of Fig. 5.2, the probability of a 1 to appear at the input of a two-input OR gate is assumed to fall within 0 to 1. Then $P(f_2 = 1) = 0–1$ and, hence, $P(f_3 = 1) = P(f_2 = 1)P(x_3 = 1) = 0–P(x_3 = 1)$. The probability of a 1 to appear at the output of the single-input ANDgate is $P(x_1 = 1)P(x_2 = 1)$, and the output probability is respectively $P(F = 1) = P(f_1 = 1) + P(f_3 = 1) - P(f_1 = 1)P(f_3 = 1)$. By substituting the upper $(P(x_3 = 1))$ and lower (0) bounds of $P(f_3 = 1)$ into the latter expression, we finally obtain

$$P(F = 1) = [P(x_1 = 1)P(x_2 = 1)] \div [P(x_1 = 1)P(x_2 = 1) + P(x_3 = 1) \\ - P(x_1 = 1)P(x_2 = 1)P(x_3 = 1)].$$

For $P(x_1 = 1) = P(x_2 = 1) = P(x_3 = 1) = p$, we obtain

$$P(F = 1) = p^2 \div (p + p^2 - p^3)$$

where p^2 and $p + p^2 - p^3$ are respectively the lower and the upper probability estimates $P(F = 1)$.

More complex is the problem of probabilistic description of sequential circuits. Analytic approaches are impracticable here. The only choice is the use of statistical experiment. An example of the implementation of the statistical approach for probabilistic description of digital circuits is STAFAN (STAtistical Fault ANalysis) [77]. It is a software system that allows one, in particular, to estimate the probability of a 1 to appear at any of the circuit nodes.

A probabilistic description of a digital circuit that has been obtained either analytically or from experimental data is required only for generation of reference probabilities to be used for comparison with the actual estimates obtained on teting the digital circuit. Such probabilities are often referred to as signal probabilities [75, 76]. They can be used to estimate the probability of a 1 to appear at the specified nodes of the circuit under test.

The problem of estimating the probability of fault detection at the specified nodes of a digital circuit is most important since it makes it possible to evaluate the efficiency of probabilistic testing in each particular case and outline the methods of improvement. Consider some practical approaches used to evaluate the probability of specified fault detection.

5.3. FAULT DETECTION PROBABILITY ESTIMATION

The probability $P_d(f \equiv \kappa)$, $\kappa \in \{0, 1\}$, of detecting a stuck-at fault $\equiv \kappa$ on node f is a measure of random testing efficiency and characterizes the complexity of defining the specified circuit fault. To comprehend the fault detection probability concept, consider the digital circuit implemented as an n-input AND gate.

As for the general case, the procedure testing for the AND gate consists of applying sequences of random bits with probabilities $P(x_1 = 1)$, $P(x_2 = 1), \ldots, P(x_n = 1)$ to its inputs and analysing its output response. Without loss of generality, suppose that $P(x_1 = 1) = P(x_2 = 1) = \cdots = P(x_n = 1) = p$.

The probability $P_d(f \equiv \kappa)$ of detecting a fault $\equiv \kappa$ is characterized by the probability of two concurrent events: fault manifestation at its point of origin and its propagation to one of the circuit outputs. The manifestation of fault κ can be estimated by probability $P_e(f \equiv \kappa)$ of its manifestation at the specified node f; the fault propagation can be estimated by probability $P_t(f \equiv \kappa)$ of its observation at any circuit output.

For an n-input AND gate, the probability of fault $\equiv 0$ to occur at the primary input will be the same as the probability of a 1 to appear at the same input, i.e. $P_e(x_1 = 0) = P(x_1 = 1) = p$, and, hence, $P_e(x_1 = 1)$ are the probabilities of a 0 to appear at the specified input. Then $P_e(x_1 = 1) = P(x_1 = 0) = 1 - P(x_1 = 1) = 1 - p$. A propagation condition for any fault at the primary input to an AND gate is path sensitization from the specified input to the gate output. The above condition is determined by the event $x_2 = 1, x_3 = 1, \ldots, x_n = 1$ whose probability is calculated as $P(x_2 = 1, x_3 = 1, \ldots, x_n = 1)$. Allowing for the independence of input random sequences, we finally obtain $P_t(x_1 \equiv \kappa) = P(x_2 = 1)P(x_3 = 1) \cdots P(x_n = 1) = p^{n-1}$. The probability $P_e(y \equiv \kappa)$ of a fault κ to appear at the AND gate output is $P(y = 1)$ for $\kappa \equiv 0$ and $P(y = 0)$ for $\kappa \equiv 1$, which the probability $P_t(y \equiv \kappa)$ being rigorously 1 since the AND gate output is the output node of the circuit.

For the straightforward structure of the digital circuit just discussed, fault manifestation and propagation are independent events and therefore the probability $P_d(f \equiv \kappa)$ may be calculated by the following relation

$$P_d(f \equiv \kappa) = P_e(f \equiv \kappa)P_t(f \equiv \kappa). \tag{5.7}$$

Table 5.1 gives the probabilities $P_e(f \equiv \kappa)$, $P_t(f \equiv \kappa)$ and $P_d(f \equiv \kappa)$ for stuck-at faults at the ith input ($i = 1$ to n) and output of AND and OR gates. Thus for $p = \frac{1}{2}$, by the relations of Table 5.1, the probability $P_d(x_i \equiv 0)$ for a four-input AND gate is $p = 1/2^4$ and $P_d(y \equiv 1) = 1 - 1/2^4$.

The concepts of fault detection, manifestation and propagation probabilities are to a great extent similar to those of controllability, observability and testability of a digital circuit. The only difference is that the newly introduced attributes of detection,

Table 5.1.

| $f \equiv \kappa$ | AND | | | OR | | |
	P_e	P_t	P_d	P_e	P_t	P_d
$x_i \equiv 1$	$1 - p$	p^{n-1}	$p^{n-1} - p^n$	$1 - p$	$(1-p)^{n-1}$	$(1-p)^n$
$x_i \equiv 0$	p	p^{n-1}	p^n	p	$(1-p)^{n-1}$	$p(1-p)^{n-1}$
$y \equiv 1$	$1 - p^n$	1	$1 - p^n$	$(1-p)^n$	1	$(1-p)^n$
$y \equiv 0$	p^n	1	p^n	$1 - (1-p)^n$	1	$1 - (1-p)^n$

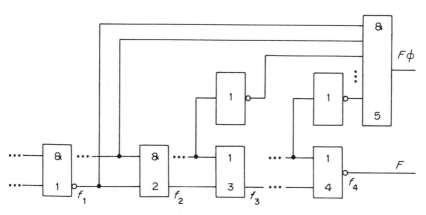

Fig. 5.3. Estimation of $f_1 \equiv 0$ detection probability reduced to estimating the signal probability at the output $F\varphi$ of the simulated AND gate

manifestation and propagation probabilities are calculated both for a uniform distribution of test patterns and for a random distribution of input probabilities.

Analytical estimation of probabilities P_d, P_e and P_t for a typical digital circuit becomes more complicated due to dependence of the fault manifestation and propagation events on one of the circuit outputs. Therefore, in the majority of cases, only a fault detection probability is estimated. Thus, in [76], the problem of the specified probability estimation reduces to calculation of a signal probability at the output of the simulated AND gate (Fig. 5.3.). The digital circuit shown in Fig. 5.3 consists of series-connected gates NAND (1), AND (2), OR (3) and NOR (4).

Let the fault be a stuck-at fault $f_1 \equiv 0$, which propagates to the output along the path through gates 2–4. Fault manifestation at its origin is conditioned by a 1 signal at the output of the first gate; fault propagation to the output is conditioned by a 1 appearing at the primary input to gate 2 and 0s at the primary inputs to gates 3 and 4. The given set of conditions that determine fault observability at the circuit output may be united by inserting a dummy AND gate 5 whose output value will be 1 if and only if all the previous conditions have been satisfied. Hence, the probability $P(F_\varphi = 1)$ of the output value to be 1 for the dummy AND gate will be equal to that of the specified fault detection

$$P_d(f_1 \equiv 0) = P(F_\varphi = 1)$$

which implies that the $f_1 \equiv 0$ detection probability corresponds to the signal probability of node F_φ. Thus fault detection probability estimation reduces to finding a signal probability. Any of the techniques discussed earlier may be used for the purpose.

When there are many fault propagation paths to the digital circuit outputs, the signal probability for the output node of a dummy AND gate will be used as a lower bound to the desired fault detection probability [76]. Also, this eliminates the need to estimate the detection probability for all possible faults in the circuit. It will suffice to calculate the detection probability for representative faults, i.e. faults that lie on the input lines to the circuit and those associated with fan-out points [76]. Such faults are illustrated in Fig. 5.4 and summarized in Table 5.2. The table gives an

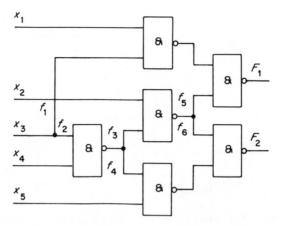

Fig. 5.4. A combinational circuit for which representative fault probabilities are estimated for nodes $x_1, x_2, x_3, x_4, x_5, f_1, f_2, f_3, f_4, f_5$ and f_6

Table 5.2.

Fault	P_d	P_d^*	Fault	P_d	P_d^*
$x_1 \equiv 0$	3/16	5/32	$f_1 \equiv 0$	3/16	5/32
$x_1 \equiv 1$	3/16	5/32	$f_1 \equiv 1$	1/8	1/8
$x_2 \equiv 0$	11/32	9/32	$f_2 \equiv 0$	3/16	5/64
$x_2 \equiv 1$	11/32	9/32	$f_2 \equiv 1$	3/16	1/16
$x_3 \equiv 0$	9/32	1/16	$f_3 \equiv 0$	11/32	3/16
$x_3 \equiv 1$	9/32	1/16	$f_3 \equiv 1$	1/8	5/64
$x_4 \equiv 0$	3/16	5/64	$f_4 \equiv 0$	3/16	3/16
$x_4 \equiv 1$	3/16	1/16	$f_4 \equiv 1$	1/8	5/64
$x_5 \equiv 0$	3/16	3/16	$f_5 \equiv 0$	7/16	3/8
$x_5 \equiv 1$	3/16	3/16	$f_5 \equiv 1$	5/16	9/32
			$f_6 \equiv 0$	7/16	25/64
			$f_6 \equiv 1$	3/16	3/16

accurate estimate of fault detection probability P_d and its lower bound P_d^* obtained as a signal probability for the dummy AND gate for each case.

The analysis of fault detection probabilities makes it possible to estimate the effectiveness of random testing for a digital circuit as well as to determine the minimal length of the test sequence.

5.4. TEST SEQUENCE LENGTH CALCULATION

A major parameter of any testing and diagnosing system for digital circuits is the time consumed by the testing procedure, which is defined uniquely by a test sequence length. for deterministic test design methods, the test length is determined by the required coverage for all possible faults in the circuit and does not exceed 1000 patterns on average. Note that fault coverage increases with number of test patterns. A similar

Table 5.3.

Circuit type	No. of inputs	No. of gates	Fault coverage(%) for random test patterns			Fault coverage(%) for deterministic test pattern
			100	1000	10000	
LSI-1	63	926	86.1	94.1	96.3	96.7
LSI-2	54	1103	75.2	92.3	95.9	97.1

relationship is evidently true for random testing as well. This is demonstrated by the example of using random test patterns for testing two LSI chips [78]. As seen from Table 5.3 [78], fault coverage for both LSI chips increases with number of test patterns, which are uniformly distributed random numbers.

The relationship between fault coverage and number of test patterns represented by random numbers is shown in Fig. 5.5. The figure implies that the fault coverage T achieved by generation of N random test patterns can be approximated as a function of an exponent [79]:

$$T = [1 - \exp(-\lambda \log_{10} N)] \times 100\% \tag{5.8}$$

where λ is a constant that assumes a specific value for each circuit type. A similar relationship may be obtained by arguing on the basis of the fault whose detection probability is P_d, with P_d in this case characterizing the probability of detection of the specified fault by applying only one test pattern. With N distinct random patterns applied, the indicated probability will increase and in the general case will be defined by

$$P_d(N) = 1 - P_n(N) \tag{5.9}$$

where $P_d(N)$ is the probability of fault detection by applying N test patterns and $P_n(N)$ is the non-detection probability of the fault specified. Considering that the

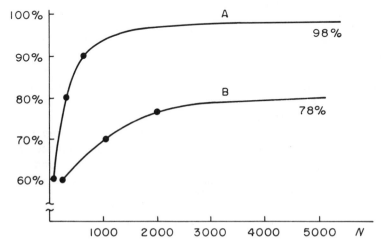

Fig. 5.5. Fault coverage versus test pattern length: A, for programmable logic array 1 (PLA-1); B, for PLA-2

generated test patterns are independent, we may calculate probability $P_n(N)$ as the product of N terms $(1 - P_d)$ representing the non-detection probability for a fault by applying a single test pattern. Subject to the latter, the relation (5.9) reduces to

$$P_d(N) = 1 - (1 - P_d)^N$$

whence we can find the number of random test patterns for the specified test validity defined by $P_d(N)$ as

$$N = \frac{\ln [1 - P_d(N)]}{\ln (1 - P_d)}. \tag{5.10}$$

For an n-input AND gate, the number of test patterns required to detect its fault $y \equiv 0$ (see Table 5.1) with probability $P_d(N) = 0.99$ may be calculated by equation (5.10) subject to the equality $P_d = p^n$:

$$N \simeq \ln (0.01)/\ln (1 - p^n)$$

where p is the probability of a 1 to appear in random patterns applied to the inputs to the AND gate.

The analysis of the latter relation shows that the test sequence length N depends heavily on the form of the circuit under test (here on the number n of AND gate inputs) and on the input probability distribution p. The example discussed in [75] for the digital circuit employing NAND gates shows that, with input probability $p = 0.5$, a 13-level circuit can be tested with probability $P_d(N) = 0.99$ by test sequences consisting of $N = 3 \times 10^8$ patterns. The number of required test patterns N reduces to 2×10^4 when the specified input probabilities are 0.617.

The above examples are indicative of the need to solve the problem of estimating input probabilities that allow one to minimize the test sequence length.

5.5. METHODS FOR OPTIMAL SELECTION OF INPUT VARIABLE PROBABILITIES

The optimal probability of input variables is that which allows one to obtain the maximal length N of test sequence to ensure the specified detection probability for the worst-case fault [80]. The worst-case fault is that which has the minimal value of its detection probability P_d. Thus the major problem to be solved in finding the optimal probabilities for input variables is to ensure the maximal possible value of the minimal fault detection probability in the circuit.

As an example demonstrating the estimation of optimal values for input probabilities, consider an n-input AND gate whose fault detection probabilities are given in Table 5.1. Probabilities $P_d(x_i \equiv 1) = p^{n-1} - p^n, P_d(y \equiv 1) = 1 - p^n$ and $P_d(x_i \equiv 0) = P_d(y \equiv 0) = p^n$ plotted for $n = 2$ and 3 are shown in Figs 5.6 and 5.7. The analysis of dependences between P_d and p shows that

$$\max_p \min (p^{n-1} - p^n, 1 - p^n, p^n)$$

for $n = 2$ is obtained with $p = 0.5$ and for $n = 3$ with $p = 2/3 = 0.666$. For a random n, we may define the optimal probability p_o of a 1 to appear at the inputs to the

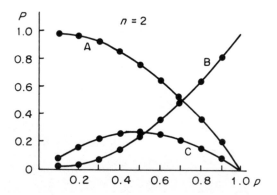

Fig. 5.6. Fault detection probability for a two-input AND gate: A, $P_d = 1 - p^2$; B, $P_d = p^2$; C, $P_d = p - p^2$

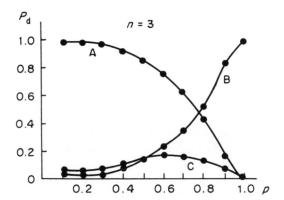

Fig. 5.7. Fault detection probability as a function of input probability p for a three-input AND gate: A, $P_d = 1 - p^3$; B, $P_d = p^3$; C, $P_d = p^2 - p^3$

AND gate by

$$\delta(p^{n-1} - p^n)/\delta p = 0$$

whence we obtain $p_o = (n - 1)/n$. For a digital circuit employing n-input OR gates, we may find p_o in a similar manner. Here, p_o equals $p = 1 - p^n$. Hence $p_o = 0.617$ with $n = 2$.

It should be noted, however, that the problem of selecting the optimal input probabilities for the general case is a linear optimization problem, which requires complex calculations [80].

A method for selecting the optimal values of input probabilities for random testing is suggested in [80]. The idea behind the method is in maximizing the minimal detection probability for those circuit faults that are most difficult to detect (faults with minimal detection probabilities). It has been shown that the input lines to the circuit as well as lines at the origin of fan-out points are the locations of faults with minimal detection probability. The optimal assignment of input vector probability is based on maximizing the objective, which represents the minimum fault detection probability. For this purpose, the so-called relaxation method has been used, but its

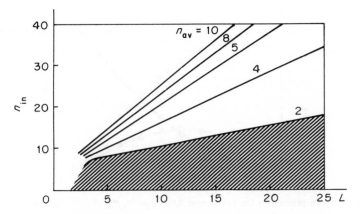

Fig. 5.8. Graphic chart showing the effective areas for random testing: n_{in} is the number of inputs to the circuit; L is the number of series-connected gates between circuit inputs and outputs; n_{av} is the average number of inputs to the circuits

assignment is not always optimal. However, the assignment obtained by the said method is always better than in the trivial case when input probabilities are 0.5.

A similar problem of optimal selection of input probability distributions has been presented in [81], where the case for the combinational circuit is also given. It has been shown that for any digital circuit comprising memory elements the problem of calculating optimal input probabilities is more complex than that of generating deterministic test sequences. For this reason, the application of random techniques proves to be inefficient.

A way of estimating the efficiency of random testing for digital circuits on minimal information has been presented in [82]. Whether random testing is suitable may be determined by the three basic parameters of the circuit: the number of inputs n_{in} to the circuit, the maximal number L of gates that have been series-connected between the inputs and outputs of the circuit, and the average number n_{av} of inputs to the circuit gates. In spite of the simplified approach that has been used in [82], the schematic representation of reported results (Fig. 5.8) may be useful for elaborating practical recommendations as to the use of random testing. In Fig. 5.8, the hatched area represents the parameters for which the random testing efficiency is very low.

The use of pseudorandom test sequences, from which pseudorandom testing of digital circuits originated, may be considered as a further development of random testing.

6

PSEUDORANDOM TESTING

6.1. PSEUDORANDOM SEQUENCES

Two methods are most commonly used for generating pseudorandom sequences. The first method, which forms the basis of a major proportion of present-day program sensors, makes use of the recurrence relation [47, 83]:

$$X_k = AX_{k-1} + B(\text{mod } M) \qquad k = 1, 2, 3, \ldots \qquad (6.1)$$

where A, B, M are constants and $X_0 > 0$, $A > 0$, $B \geqslant 0$, $M > X_0$, $M > A$, $M > B$. This method has been called multiplicative congruential (with $B = 0$) or mixed congruential (with $B > 0$), so a sequence generated by the method is called linear congruential. With an appropriate choice of the value of $M = r^m$, where r is the basis number that is used to represent the values of m-digit pseudorandom numbers X_k, equation (6.1) may be rearranged to give

$$X_k = AX_{k-1} + B(\text{mod } r^m) \qquad k = 1, 2, 3, \ldots .$$

Not all sets of values X_0, A, B, r and m give rise to sequences whose properties are close to those of uniformly distributed random sequences [84]. For example, with $X_0 = 110$, $A = 011$, $B = 001$, $r = 2$ and $m = 3$, the generated number sequence has the form

$$X_0 = 110$$
$$X_1 = 011$$
$$X_2 = 010$$
$$X_3 = 111$$
$$X_4 = 110.$$

As is evident, the values of the low-order digit in the pseudorandom number sequence $\{X_k\}$ ($\{ X_k\} = X_0, X_1, X_2, \ldots$) have period 2. For the high-order digit and the digit generated in the second bit, the periods are 4 and 1, respectively. For the general case of [47], the period L_γ for a γth digit, $\gamma \in \{1, 2, 3, \ldots, m\}$, in the pseudorandom number sequence $\{X_k\}$ can be estimated by

$$\max L_\gamma = r^{\gamma - 1}(r - 1) \qquad (6.2)$$

with $B \neq 0$ or by

$$\max L_\gamma = r^{2\gamma - 1}(r - 1) \qquad (6.3)$$

with $B \neq 0$.

Thus, for binary representation ($r = 2$), we find that the period L_1 of the low-order digit is 1 by (6.2). Hence the low-order digit in a random sequence $\{X_k\}$ is invariable and equal to 1. The second digit period $L_2 \leqslant 2$, thereby denoting that its value is either invariable or reversible in each cycle.

By using equation (6.1) with $B \neq 0$, the period for digits in the numbers can be slightly increased as follows from equation (6.3). However, the above periodicity cannot be radically eliminated, thereby affecting the quality of pseudorandom number sequence [84]. As applied to the problem of digital circuit testing, periodicity can significantly degrade the test coverage. Thus, for example, by applying a constant logical signal to a definite input of a digital circuit it is not possible to detect even a stuck-at fault associated with the generated signal value at the tested input node. What is more, implementation of the discussed pseudorandom number generation method and its modifications is very complicated due to multiplications involved. Because the above method exhibits the mentioned deficiencies, the second method is more popular in generating pseudorandom sequences by means of the relation [47]

$$a_k = \sum_{i=1}^{m} \alpha_i a_{k-1} \qquad k = 0, 1, 2, \ldots \qquad (6.4)$$

where k is the clock number, $a_k \in \{0, 1\}$ are the sequence digits, $\alpha_i \in \{0, 1\}$ are constant coefficients and $\sum_{i=1}^{m}$ represents modulo-2 addition of m logic variables. With an appropriate choice of the α_i coefficients, the bit sequence $\{a_k\}$ has maximal length $2^m - 1$ by the characteristic polynomial $\varphi(x) = 1 \oplus \alpha_1 x^1 \oplus \alpha_2 x^2 \oplus \cdots \oplus \alpha_{m-1} x^{m-1} \oplus \alpha_m x^m$, which must be primitive. Such a sequence is called an M-sequence [47]. Sec. 3.5 deals with the case when $\varphi(x) = 1 \oplus x^3 \oplus x^4$, and Fig. 3.16(a) shows its block diagram, the associated memory elements of which assume the values of Table 6.1. the table shows that the period for values of the ith digit $a_i(k)$, $i = 1$ to 4, of an M-sequence generator is $2^4 - 1 = 15$.

In synthesizing an M-sequence generator a major problem is to find a polynomial $\varphi(x)$ that meets the conditions for maximal-length sequence generation. It is known that for a given m-deg $\varphi(x)$ there exist $\Phi(L = 2^m - 1)/m$ distinct primitive polynomials where $\Phi(L)$ is the Euler function [72].

Since $\Phi(L)$ increases quickly with m, the number of polynomials $\varphi(x)$ of degree m that generate M-sequences will accordingly increase. Among the set of polynomials

Table 6.1.

k	$a_1(k)$	$a_2(k)$	$a_3(k)$	$a_4(k)$	k	$a_1(k)$	$a_2(k)$	$a_3(k)$	$a_4(k)$
0	1	0	0	0	9	1	1	0	1
1	0	1	0	0	10	1	1	1	0
2	0	0	1	0	11	1	1	1	1
3	1	0	0	1	12	0	1	1	1
4	1	1	0	0	13	0	0	1	1
5	0	1	1	0	14	0	0	0	1
6	1	0	1	1	15	1	0	0	0
7	0	1	0	1	16	0	1	0	0
8	1	0	1	0	17	...	...	...	...

Table 6.2.

m	$\varphi(x)$	m	$\varphi(x)$
1	$1 \oplus x$	21	$1 \oplus x^2 \oplus x^{21}$
2	$1 \oplus x \oplus x^2$	22	$1 \oplus x \oplus x^{22}$
3	$1 \oplus x \oplus x^3$	23	$1 \oplus x^5 \oplus x^{23}$
4	$1 \oplus x \oplus x^4$	24	$1 \oplus x^3 \oplus x^4 \oplus x^{24}$
5	$1 \oplus x^2 \oplus x^5$	25	$1 \oplus x^3 \oplus x^{25}$
6	$1 \oplus x \oplus x^6$	26	$1 \oplus x \oplus x^7 \oplus x^8 \oplus x^{26}$
7	$1 \oplus x \oplus x^7$	27	$1 \oplus x \oplus x^7 \oplus x^8 \oplus x^{27}$
8	$1 \oplus x \oplus x^5 \oplus x^6 \oplus x^8$	28	$1 \oplus x^3 \oplus x^{28}$
9	$1 \oplus x^4 \oplus x^9$	29	$1 \oplus x^2 \oplus x^{29}$
10	$1 \oplus x^3 \oplus x^{10}$	30	$1 \oplus x \oplus x^{15} \oplus x^{16} \oplus x^{30}$
11	$1 \oplus x^2 \oplus x^{11}$	31	$1 \oplus x^3 \oplus x^{31}$
12	$1 \oplus x^3 \oplus x^4 \oplus x^7 \oplus x^{12}$	32	$1 \oplus x \oplus x^{27} \oplus x^{28} \oplus x^{32}$
13	$1 \oplus x \oplus x^3 \oplus x^4 \oplus x^{13}$	33	$1 \oplus x^{13} \oplus x^{33}$
14	$1 \oplus x \oplus x^{11} \oplus x^{12} \oplus x^{14}$	34	$1 \oplus x \oplus x^{14} \oplus x^{15} \oplus x^{34}$
15	$1 \oplus x \oplus x^{15}$	35	$1 \oplus x^2 \oplus x^{35}$
16	$1 \oplus x^2 \oplus x^3 \oplus x^5 \oplus x^{16}$	36	$1 \oplus x^{11} \oplus x^{36}$
17	$1 \oplus x^3 \oplus x^{17}$	37	$1 \oplus x^2 \oplus x^{10} \oplus x^{12} \oplus x^{37}$
18	$1 \oplus x^7 \oplus x^{18}$	38	$1 \oplus x \oplus x^5 \oplus x^6 \oplus x^{38}$
19	$1 \oplus x \oplus x^5 \oplus x^6 \oplus x^{19}$	39	$1 \oplus x^4 \oplus x^{39}$
20	$1 \oplus x^3 \oplus x^{20}$	40	$1 \oplus x^2 \oplus x^{19} \oplus x^{21} \oplus x^{40}$

$\varphi(x)$ with identical high degree m there may exist polynomials with the least number of unit coefficients α_i, $i = 1$ to m. This condition is characteristic for the structures based on M-sequences and most simple in implementation. Table 6.2 shows polynomials with a minimal number of non-zero coefficients for all $m \leqslant 40$ [47]. A full table of irreducible primitive polynomials up to degree $m = 34$ is given in [85]. A more comprehensive collection of polynomials classified according to different criteria is given in [86–88].

Note that for generating M-sequences a polynomial $\varphi^{-1}(x) = x^m \varphi(x^{-1})$, which is the reverse of the primitive polynomial $\varphi(x)$, may also be used. The resulting maximal-length sequence will be the reverse of the sequence generated by $\varphi(x)$ [89]. For example, polynomial $\varphi^{-1}(x) = x^4(1 \oplus x^{-3} \oplus x^{-4}) = 1 \oplus x^1 \oplus x^4$ is the reverse of the polynomial $\varphi(x) = 1 \oplus x^3 \oplus x^4$.

A major advantage of the pseudorandom sequence generation method as compared to that of (6.4) is the simplicity of its implementation, hardware implementation included.

The M-sequence generator that functions in accordance with expression (6.4) comprises only an m-bit shift register (SR) and the set of modulo-2 adders in the feedback circuit. In operation the SR stores m-bits of the M-sequence and shifts the m-bit code one bit to the right. The modulo-2 adders in the feedback circuit compute the values of successive bits that are sequentially written to the left-most SR position [47]. The SR states can be represented as a sequence of m-dimensional vectors $A(k) = a_1(k)a_2(k)a_3(k)\cdots a_m(k)$, $k = 0, 1, 2, ...$, where $a_i(k) \in \{0, 1\}$ with $i = 1$ to m, for which the following relation applies [72].

$$
\begin{vmatrix} a_1(k) \\ a_2(k) \\ a_3(k) \\ \vdots \\ a_m(k) \end{vmatrix} = \begin{vmatrix} \alpha_1 & \alpha_2 & \alpha_3 & \cdots & \alpha_{m-1} & \alpha_m \\ 1 & 0 & 0 & \cdots & 0 & 0 \\ 0 & 1 & 0 & \cdots & 0 & 0 \\ \vdots & \vdots & \vdots & & \vdots & \vdots \\ 0 & 0 & 0 & \cdots & 1 & 0 \end{vmatrix} \begin{vmatrix} a_1(k-1) \\ a_2(k-1) \\ a_3(k-1) \\ \vdots \\ a_m(k-1) \end{vmatrix}.
$$

More compactly, $A(k) = V^1(A(k-1)$, from whence the relation

$$
A(k+s) = V^{s+1}A(k-1) \tag{6.5}
$$

where

$$
V = \begin{vmatrix} \alpha_1 & \alpha_2 & \alpha_3 & \cdots & \alpha_{m-1} & \alpha_m \\ 1 & 0 & 0 & \cdots & 0 & 0 \\ 0 & 1 & 0 & \cdots & 0 & 0 \\ \vdots & \vdots & \vdots & & \vdots & \vdots \\ 0 & 0 & 0 & \cdots & 1 & 0 \end{vmatrix}
$$

holds for any s.

For a specific M-sequence defined by polynomial $\varphi(x) = 1 \oplus x^3 \oplus x^4$, the matrix V has the form

$$
V = \begin{vmatrix} 0 & 0 & 1 & 1 \\ 1 & 0 & 0 & 0 \\ 0 & 1 & 0 & 0 \\ 0 & 0 & 1 & 0 \end{vmatrix}.
$$

By sequentially applying equation (6.4) or (6.5) for $s = 4$ we may generate single- or multiple-bit pseudorandom number sequences that have some statistical properties.

Consider now the most significant properties of maximal-length sequences (M-sequences).

(i) The period of an M-sequence generated by (6.4) is described by the high degree of generating polynomial $\varphi(x)$ and is $L = 2^m - 1$.

(ii) For the specified polynomial $\varphi(x)$, there exist L distinct M-sequences that differ by phase shifts. Thus 15 M-sequences (Table 6.3) are associated with the polynomial $\varphi(x) = 1 \oplus x^3 \oplus x^4$ [47].

(iii) In an M-sequence a_k, $k = 0, 1, 2, ..., L-1$, there are 2^{m-1} 1s and $2^{m-1} - 1$ 0s. Their probabilities of occurrence are given by [47]:

$$
P(a_k = 1) = \frac{2^{m-1}}{2^m - 1} = \frac{1}{2} + \frac{1}{2^{m+1} - 2}
$$

$$
P(a_k = 0) = \frac{2^{m-1} - 1}{2^m - 1} = \frac{1}{2} - \frac{1}{2^{m+1} - 2}
$$

and can take on any values, however close to $\frac{1}{2}$, as m increases.

(iv) In an M-sequence, the probability of runs of L, $L \in \{1, 2, ..., m-1\}$, identical digits (1s or 0s) closely approximates that of the random sequence [72].

(v) For any s $(1 \leqslant s < L)$ there exists an $r \neq s$ $(1 \leqslant r < L)$ such that $\{a_k\} \oplus \{a_{k-s}\} = \{a_{k-r}\}$. This is commonly called the 'shift-and-add' property [24].

Table 6.3.

							{a_i}							
0	0	0	1	0	0	1	1	0	1	0	1	1	1	1
0	0	1	0	0	1	1	0	1	0	1	1	1	1	0
0	1	0	0	1	1	0	1	0	1	1	1	1	0	0
1	0	0	1	1	0	1	0	1	1	1	1	0	0	0
0	0	1	1	0	1	0	1	1	1	1	0	0	0	1
0	1	1	0	1	0	1	1	1	1	0	0	0	1	0
1	1	0	1	0	1	1	1	1	0	0	0	1	0	0
1	0	1	0	1	1	1	1	0	0	0	1	0	0	1
0	1	0	1	1	1	1	0	0	0	1	0	0	1	1
1	0	1	1	1	1	0	0	0	1	0	0	1	1	0
0	1	1	1	1	0	0	0	1	0	0	1	1	0	1
1	1	1	1	0	0	0	1	0	0	1	1	0	1	0
1	1	1	0	0	0	1	0	0	1	1	0	1	0	1
1	1	0	0	0	1	0	0	1	1	0	1	0	1	1
1	0	0	0	1	0	0	1	1	0	1	0	1	1	1

(vi) The M-sequence autocorrelation function is defined [72] by

$$R_a(\tau) = \begin{cases} 1 & \text{for } \tau = 0 \, (\text{mod } L) \\ -1/L & \text{for } \tau \neq 0 \, (\text{mod } L). \end{cases}$$

(vii) Among the L non-zero M-sequences formed by the polynomial $\varphi(x)$ there exists a single sequence with the property $a_k = a_{2k}$, $k = 0, 1, 2, \ldots$, which is called characteristic [90] and defined as follows. For the specified polynomial $\varphi(x)$, the system of linear equations

$$a_i = a_{2i} \qquad i = 0 \text{ to } m - 1 \qquad (6.6)$$

is written in accordance with the recurrence relation (6.4). The unique non-zero solution of the system (6.6) will be the characteristic sequence associated with the given $\varphi(x)$.

By way of example, we shall find a characteristic sequence for the generating polynomial $\varphi(x) = 1 \oplus x^3 \oplus x^4$ [47]. For this case the system (6.6) takes the form

$$a_0 = a_0,$$
$$a_1 = a_2,$$
$$a_2 = a_4 = a_0 \oplus a_1,$$
$$a_3 = a_6 = a_2 \oplus a_3.$$

By solving the system of equations, we shall find that the initial values of the characteristic sequence $\{a_k\}^*$ associated with the generating polynomial $\varphi(x) = 1 \oplus x^3 \oplus x^4$ are defined by the relation $a_0 a_1 a_2 a_3 = 1000$.

(viii) Decimation of the M-sequence $\{a_i\}$ by q ($q = 1, 2, 3, \ldots$) means the generation of another sequence $\{b_k\}$ of qth elements of $\{a_k\}$, i.e. $b_k = a_k$ [89]. If $\{b_k\}$ is a non-zero sequence it is generated by polynomial $\varphi'(x)$ whose roots are the qth degrees of the original polynomial $\varphi(x)$ roots and its period is $L/(L, q)$, where (L, q) is the least common divisor of L and q. With $(L, q) = 1$, the period of $\{b_k\}$ is $2^m - 1$, where $m = \deg$

$\varphi(x)$, and decimation is said to be proper or normal [47]. Any proper decimation results in an M-sequence generated by a polynomial $\varphi'(x)$. In this case, decimation performed on a sequence that has been shifted by s clocks relative to an original $\{a_k\}$ will result in a sequence shifted by several clocks relative to $\{b_k\}$. In other words, irrespective of the shift value that has been chosen for $\{a_k\}$, decimation will always result in an M-sequence generated by polynomial $\varphi'(x)$ [47].

As a final note, it is worth mentioning that the theory of M-sequence generation and M-sequence properties are detailed in [89, 91, 92].

6.2. PSEUDORANDOM SEQUENCES IN THE TESTING OF DIGITAL CIRCUITS

When applying random number sequences as test patterns, the digital circuit (DC) test coverage can be expressed by the value of $P_d(N)$ [93], which is the fault detection probability determined as a ratio of detected faults to the total number of DC faults on using N test patterns. The probability $P_d(N)$, which does not decrease with N, may be expressed as

$$P_d(N) = 1 - P_n(N)$$

when $P_n(N)$ is a 'fault escape' probability function, which does not increase with increasing N.

For random test patterns and non-redundant DC the following relations hold. With $N = 0$, $P_n(N) = 1$, as $N \to \infty$, $\lim P_n(N) = 0$ and hence $P_d(N) = 0$ for $N = 0$ and $P_d(N) = 1$ for $N = \infty$. At the same time for pseudorandom test patterns (PRTPs) the following limiting relation holds

$$\lim_{N \to \infty} P_n(N) = p_1 \qquad (6.7)$$

where $0 \leqslant p_1 \leqslant 1$ is a constant determined by the form of pseudorandom sequences used as well as by the topology of the DC under test. Thus for a combination DC the value of p_1 may differ from 0 and is determined as the ratio of undetected faults to their total number. Condition $p_1 \neq 0$ is met due to the deterministic nature of the PRTP. Let us consider how it manifests itself in a specific case. Fig. 6.1 shows a portion of a sequential DC. The latter is tested by a PRTP generator, which generates an M-sequence by polynomial $\varphi(x) = 1 \oplus x^3 \oplus x^4$. As evident from the figure, a $\equiv 0$ fault on the output of the EOR gate, a $\equiv 1$ on the output of the OR gate and a bridging fault on the inputs of the AND gate are not detected, since both inputs of the gates in question are applied with identical logic signals. The appearance of untestable portions in a sequential DC is detailed in [47, 94]. It has been proved that for any PRTP generator there can be found a DC that is untestable on inputs of some of its gates. Condition $p_1 = 0$ can be met by designing a PRTP generator for the specified DC or a class of such DCs.

We shall consider the requirements for hardware pseudorandom test pattern generators (PTPGs) that satisfy the condition $p_1 = 0$ when testing a DC that is a sequential synchronous circuit.

Any sequential circuit may be represented by the so-called Huffman model, for

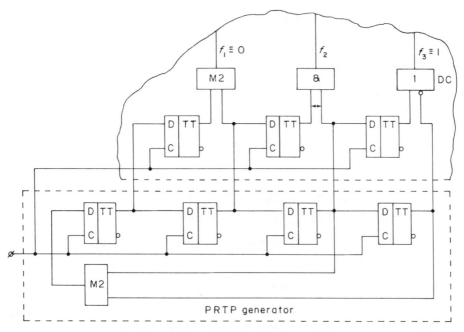

Fig. 6.1. An example of a digital circuit

which $A(k) = a_1(k), \ldots, a_c(k), \ldots, a_c(k)$ and $Y(k) = y_1(k), \ldots, y_r(k), \ldots, y_R(k)$ are, respectively, its input and output variables in the kth cycle of operation, and $\mathbf{Z}(k) = z_1(k), \ldots,$ $z_q(k), \ldots, z_Q(k)$ are internal variables that define the circuit state.

The circuit behaviour is described by the set of equations [95]

$$
\begin{aligned}
y_r(k) &= f_r[a_1(k), \ldots, a_c(k), z_1(k), \ldots, z_Q(k)] \\
&= f_r[A(k), \mathbf{Z}(k)] \qquad r = 1 \text{ to } R
\end{aligned} \tag{6.8}
$$

$$
\begin{aligned}
z_q(k + 1) &= \psi_q[a_1(k), \ldots, a_c(k), z_1(k), \ldots, z_Q(a)] \\
&= \psi_q[A(k), \mathbf{Z}(k)] \qquad q = 1 \text{ to } Q, \; k = 0, 1, 2, \ldots.
\end{aligned}
$$

Assuming that the circuit has no feedback loops and makes use of only synchronous memory elements performing the delay function, we may subdivide it into several levels of memory elements that have been functionally connected in series. Then the mathematical model of such a sequential circuit derived from the set of equations (6.8), when written in vector form, is

$$
\begin{aligned}
z_1(k + 1) &= \psi_1[A(k)] \\
z_2(k + 1) &= \psi_2[A(k), z_1(k)] \\
z_3(k + 1) &= \psi_3[A(k), z_1(k), z_2(k)] \\
&\;\;\vdots \\
z_T(k + 1) &= \psi_T[A(k), z_1(k), z_2(k), \ldots, z_{T-1}(k)] \\
Y(k) &= f[A(k), z_1(k), z_2(k), \ldots, z_T(k)]
\end{aligned} \tag{6.9}
$$

where $z_1(k) \bigcup z_2(k) \bigcup z_3(k) \bigcup \cdots \bigcup z_T(k) = \mathbf{Z}(k)$.

Thus the sequential circuit is represented as $T \leqslant Q$ levels of functionally dependent memory elements, with the state of a definite jth-level memory element ($j = 1$ to T) in the $(k+1)$th cycle of circuit operation being dependent only on the states of preceding-level memory elements in the kth cycle and the values of input variables $A(k)$.

Rearrangement of the set of equation (6.9) results in

$$z_1(k+1) = \psi_1[A(k)]$$
$$z_2(k+1) = \psi_2[A(k), A(k-1)]$$
$$z_3(k+1) = \psi_3[A(k), A(k-1), A(k-2)]$$
$$\vdots \qquad\qquad\qquad\qquad\qquad (6.10)$$
$$z_T(k+1) = \psi_T[A(k), A(k-1), A(k-2), \dots, A(k-T+1)]$$
$$Y(k) = f[A(k), A(k-1), A(k-2), \dots, A(k-T)]$$

The set of equations (6.8) may now be represented in vector form as

$$Z(k) = \psi[A(k), A(k-1), A(k-2), \dots, A(k-T+1)]$$
$$Y(k) = f[A(k), A(k-1), A(k-2), \dots, A(k-T)].$$

By using the above expressions, we may demonstrate that the value of switching function E_l on the lth node of a sequential circuit is defined as

$$F_l(k) = F_l[A(k), A(k-1), A(k-2), \dots, A(k-T)].$$

The occurrence of an untestable fault on the lth node may be attributed to the fact that $F_l(k)$ has not been tested on all possible sets of vectors $A(k), A(k-1), A(k-2), \dots,$ $A(k-T)$ at generation of input test patterns $A(k)$. Although for random test patterns $A(k)$ all possible combinations of binary vectors $A(k), A(k-1), A(k-2), \dots, A(k-T)$ can be basically derived, for PRTPs it is not the case. Thus for the circuit of Fig. 6.1 the value of f_1 is defined as $f_1(k) = f_1[A(k), A(k-1)]$ where $A(k) = a_1(k)a_2(k)a_3(k)a_4(k)$ and $A(k-1) = a_1(k-1)a_2(k-1)a_3(k-1)a_4(k-1)$. By using the relations

$$a_1(k) = a_3(k-1) \oplus a_4(k-1)$$
$$a_2(k) = a_1(k-1) \qquad a_3(k) = a_2(k-1) \qquad\qquad (6.11)$$
$$a_4(k) = a_3(k-1) \qquad k = 0, 1, 2, \dots$$

which describe the behaviour of a PRTP generator (Fig. 6.1), we obtain

$$f_1(k) = f_1[a_1(k), a_2(k), a_3(k), a_4(k), a_2(k), a_3(k), a_4(k), a_1(k) \oplus a_4(k)].$$

Thus the values of f_1, f_2 and f_3 (Fig. 6.1), which in the general case depend on eight binary variables, are not tested on all their possible combinations due to the functional dependence between $A(k)$ and $A(k-1)$ expressed by (6.11). Therefore, with the dependence between bits in patterns $A(k)$ and $A(k-1)$ causing certain combinations in the bit pattern sequence $a_1(k)a_2(k)a_3(k)a_4(k)a_1(k-1)a_2(k-1)a_3(k-1)a_4(k-1)$ to be eliminated, the given example (Fig. 6.1) displays untestable fault conditions. Exhaustive testing for the portion of a digital circuit in question is ensured by generating all possible binary combinations in two successive test patterns $A(k)$ and $A(k-1)$.

In general, it is necessary to generate all possible binary combinations in $(T+1)$

successive test patterns for a circuit containing T levels of functionally dependent sequential memory elements. This condition can be easily met by using test pattern generators based on physical concepts [94]. Such devices are impracticable in testing systems because of the need for maximum speed, rather high length and reproducibility of a test signal.

The fact that very long hardware-generated PRTPs have found extensive applications may be attributed to the above conditions. They are used in testing systems based on probability [63, 93], signature analysis [96] as well as in self-testing of LSI/VLSI circuits. A PRTP used in many cases is an M-sequence [97–102] for which the condition for generating all possible $(T+1)$ n-bit combinations in the $(T+1)$th successive test pattern will be satisfied, provided the n-bit test patterns are uniformly distributed by the $(T+1)$-dimensional distribution [95]. Then $p_1 = 0$ holds for any DC comprising not more than T functionally dependent memory levels and no feedback.

Let us consider a technique for designing PRTP generators based on M-sequences and providing full test coverage for DCs.

6.3. DESIGN OF PSEUDORANDOM TEST SEQUENCE GENERATOR

The procedure for designing hardware pseudorandom test pattern generators providing full test coverage consists of the following basic steps [47, 103].

(i) For the specified set of digital circuits, the maximum number of successive functionally dependent memory elements T and the maximum number of inputs n are determined.

(ii) From the well known tables of primitive polynomials [85], a polynomial $\varphi(x)$ is chosen that satisfies the condition $\deg \varphi(x) \geqslant (T+1)n$ and has the minimum number of non-zero coefficients $\alpha_i \in \{0, 1\}$, $i = 1$ to m.

(iii) Using the values of α_i, a matrix

$$
V = \begin{vmatrix}
\alpha_1 & \alpha_2 & \alpha_3 & \cdots & \alpha_{m-1} & \alpha_m \\
1 & 0 & 0 & \cdots & 0 & 0 \\
0 & 1 & 0 & \cdots & 0 & 0 \\
\vdots & \vdots & \vdots & & \vdots & \vdots \\
0 & 0 & 0 & \cdots & 1 & 0
\end{vmatrix}
$$

is constructed for generating the elements of matrix V^n.

(iv) From the relation (6.5) for $s = n$ the set of logic equations is written to associate the states of memory elements in the $(k + n)$th generator cycle with those in the kth cycle.

(v) Based on the resulting set of logic equations

$$A(k + n) = V^n A(k)$$

the functional diagram of the PRTP generator is constructed.

By successively executing the steps of the above algorithm for PRTP generator design, we can construct a pseudorandom number generator that implements an n-bit shift on data. Subject to that, steps (iii) and (iv) can be executed by n-fold application of the equation [104]

$$a_1(k+1) = \sum_{l=1}^{m} \alpha_l a_l(k) \tag{6.12}$$

$$a_i(k+1) = a_{i-1}(k) \qquad i = 2 \text{ to } m, \ k = 0, 1, 2, \dots.$$

By using the test of logic equations

$$a_j(k+n) = f_j[a_1(k), a_2(k), \dots, a_m(k)] \qquad j = 1 \text{ to } m \tag{6.13}$$

we can draw up a generator circuit. Besides, the connection layout for the generator being designed can be obtained with an algorithm for finding coefficients $\delta_i(n) \in \{0, 1\}, i = 1$ to m, of the linear system (6.13) [105, 106].

Consider the use of the above PRTP generator design procedure for the DC of Fig. 6.1.

(i) The number T of functionally connected DC memory elements is 1 and the number of inputs $n = 4$.

(ii) Since $(T+1)n = (1+1)4 = 8$, we choose a polynomial $\varphi(x) = 1 \oplus x^5 \oplus x^9$ (see Table 6.2) for which $m = 9$.

(iii) By using the values of coefficients $\alpha_1 = \alpha_2 = \alpha_3 = \alpha_4 = \alpha_6 = \alpha_7 = \alpha_8 = 0$ and $\alpha_5 = \alpha_9 = 1$, a matrix

$$V = \begin{vmatrix} 0 & 0 & 0 & 0 & 1 & 0 & 0 & 0 & 1 \\ 1 & 0 & 0 & 0 & 0 & 0 & 0 & 0 & 0 \\ 0 & 1 & 0 & 0 & 0 & 0 & 0 & 0 & 0 \\ 0 & 0 & 1 & 0 & 0 & 0 & 0 & 0 & 0 \\ 0 & 0 & 0 & 1 & 0 & 0 & 0 & 0 & 0 \\ 0 & 0 & 0 & 0 & 1 & 0 & 0 & 0 & 0 \\ 0 & 0 & 0 & 0 & 0 & 1 & 0 & 0 & 0 \\ 0 & 0 & 0 & 0 & 0 & 0 & 1 & 0 & 0 \\ 0 & 0 & 0 & 0 & 0 & 0 & 0 & 1 & 0 \end{vmatrix}$$

is constructed which, when used, results in

$$V^4 = \begin{vmatrix} 0 & 1 & 0 & 0 & 0 & 1 & 0 & 0 & 0 \\ 0 & 0 & 1 & 0 & 0 & 0 & 1 & 0 & 0 \\ 0 & 0 & 0 & 1 & 0 & 0 & 0 & 1 & 0 \\ 0 & 0 & 0 & 0 & 1 & 0 & 0 & 0 & 1 \\ 1 & 0 & 0 & 0 & 0 & 0 & 0 & 0 & 0 \\ 0 & 1 & 0 & 0 & 0 & 0 & 0 & 0 & 0 \\ 0 & 0 & 1 & 0 & 0 & 0 & 0 & 0 & 0 \\ 0 & 0 & 0 & 1 & 0 & 0 & 0 & 0 & 0 \\ 0 & 0 & 0 & 0 & 1 & 0 & 0 & 0 & 0 \end{vmatrix}$$

(iv) By substituting the resulting matrix into equation (6.5), we obtain the set of logic equations

$$A(k+4) = V^4 A(k).$$

In view of the single-cycle principle of PRTP generator operation, we may transform it to

$$a_1(k+1) = a_2(k) \oplus a_6(k) \qquad a_2(k+1) = a_3(k) \oplus a_7(k)$$
$$a_3(k+1) = a_4(k) \oplus a_8(k) \qquad a_4(k+1) = a_5(k) \oplus a_9(k) \qquad (6.14)$$
$$a_5(k+1) = a_1(k) \qquad a_6(k+1) = a_2(k) \qquad a_7(k+1) = a_3(k)$$
$$a_8(k+1) = a_4(k) \qquad a_9(k+1) = a_5(k).$$

(v) The functional diagram of the PRTP generator is built in accordance with the set of logic equations (6.14).

By using the values of the first four PRTP generator bits as an output test pattern $A(k)$, we can demonstrate that all possible bit combinations are generated in two successive patterns $A(k)$ and $A(k+1)$. Thus all faults in the example under consideration will be detected by the PRTP generator described by the set of equations (6.14).

Complete test coverage for a DC test involving hardware generation of pseudorandom test patterns is ensured under the following condition.

The PRTP sequence obeys the $(T+1)$-dimensional law of uniform distribution, where T is the number of levels for successive functionally connected memory elements in a DC without feedback loops. An increase in uniform distribution dimensionality makes the PRTP closer to a purely random sequence in its properties. Therefore, for a number of complex DCs with feedback loops, it is more expedient to apply PRTPs whose properties practically do not differ from those of random test patterns. Among such sequences are composite sequences consisting of algorithmically connected combinations of pseudorandom and random values [107–109]. The sequences generated by these methods manifest no periodicity, the possibility of obtaining maximal-dimension zero code, reoccurrence of any generated number in a random number of cycles, equiprobability of bits in test patterns, and low level of auto- and mutual correlation between them [47].

We consider the design of a composite PRTP generator whose operating principle is based on de Bruijn sequence randomization [94].

The de Bruijn sequence can be obtained by completing an M-sequence with a zero bit to produce an m-bit zero pattern, where m is the high degree of a generating polynomial $\varphi(x)$. By using equation (6.12), we obtain the set of logic equations describing the behaviour of the de Bruijn sequence generator

$$a_1(k+1) = \sum_{l=1}^{m} \alpha_l a_l(k) \oplus \overline{\bigvee_{l=1}^{m-1} a_l(k)} \qquad (6.15)$$

$$a_i(k+1) = a_{i-1}(k) \qquad i = 2 \text{ to } m$$

where $\alpha_i \in \{0, 1\}$, $n = 1$ to m, are constant coefficients defined by a generating polynomial $\varphi(x)$; $a_1(k) \in \{0, 1\}$ is a de Bruijn sequence element treated as the first memory element of its generator, with $a_1(k) = a_1(k+2^m)$ for $k = 0, 1, 2, \ldots$.

By the set of equations (6.15), as for the PRTP generator design, we can obtain analytic expressions for disjoint binary codes. However, in this case the above algorithm turns out to be unacceptable due to non-linear links in the system (6.15). Therefore, we shall use the technique discussed in [104] to demonstrate that the state of the mth generator bit depends on the other bit values and the number of completed cycles. Thus, for $l < m$,

$$a_m(k + l) = a_{m-1}(k + l - 1) = \cdots = a_{m-i}(k + l - i) = a_{m-l}(k).$$

Then by equation (6.15) we obtain

$$a_m(k + m) = a_1(k + 1) = \sum_{l=1}^{m} \alpha_l a_l(k) \oplus \overline{\bigvee_{l=1}^{m-1} a_l(k)} \tag{6.16}$$

in $l = m$ cycles. It can easily be shown in the same way that

$$a_{m-1}(k + m) = \alpha_1 a_1(k + 1) \oplus \sum_{l=1}^{m-1} \alpha_{l+1} a_l(k) \oplus \overline{a_1(k + 1) \vee \bigvee_{l=2}^{m-1} a_l(k + 1)}$$

$$= \alpha_1 a_m(k + m) \oplus \sum_{l=1}^{m-1} \alpha_{l+1} a_l(k) \oplus \overline{a_m(k + m) \vee \bigvee_{l=1}^{m-2} a_l(k)}. \tag{6.17}$$

For $a_{m-2}(k + m)$ we obtain

$$a_{m-2}(k + m) = \alpha_1 a_{m-1}(k + m) \oplus \alpha_2 a_m(k + m) \oplus \sum_{l=1}^{m-2} \alpha_{l+2} a_l(k)$$

$$\oplus \overline{a_m(k + m) \vee a_{m-1}(k + m) \vee \bigvee_{l=1}^{m-3} a_l(k)}. \tag{6.18}$$

Generalizing equations (6.17) and (6.18) we may write

$$a_{m-i}(k + m) = \sum_{l=1}^{i} \alpha_{i+1-l} a_{m+1-l}(k + m) \oplus \sum_{l=1}^{m-i} \alpha_{l+i} a_l(k)$$

$$\oplus \overline{\bigvee_{l=1}^{m-i-1} a_l(k) \vee \bigvee_{l=1}^{i} a_{m+1-l}(k + m)} \qquad i = 1 \text{ to } m - 2.$$

Given $i = m - 1$, we may get for $a_{m-i}(k + m)$

$$a_{m-m+1}(k + m) = a_1(k + m)$$

$$= \sum_{l=1}^{m-1} \alpha_{m-l} a_{m+1-l}(k + m) \oplus \alpha_m a_1(k) \oplus \overline{\bigvee_{l=1}^{m-1} a_{m+1-l}(k + m)}.$$

The final form of the set of equations that defines generation of an m-bit pattern of the de Bruijn sequence is

$$a_{m-i}(k + m) = \begin{cases} \displaystyle\sum_{l=1}^{m} \alpha_l a_l(k) \oplus \overline{\bigvee_{l=1}^{m-1} a_l(k)} & i = 0 \\[4ex] \displaystyle\sum_{l=1}^{i} \alpha_{i+1-l} a_{m+1-l}(k + m) \oplus \sum_{l=1}^{m-i} \alpha_{l+i} a_l(k) \\[2ex] \displaystyle\quad \oplus \overline{\bigvee_{l=1}^{m-i-1} a_l(k) \vee \bigvee_{l=1}^{i} a_{m+1-l}(k + m)} & i = 1 \text{ to } m - 2 \\[4ex] \displaystyle\sum_{l=1}^{m-1} \alpha_{m-l} a_{m+1-l}(k + m) \oplus \alpha_m a_1(k) \oplus \overline{\bigvee_{l=1}^{m-1} a_{m+1-l}(k + m)} & i = m-1. \end{cases} \tag{6.19}$$

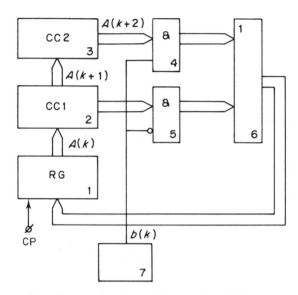

Fig. 6.2. Block diagram of a composite sequence generator: 1, static register; 2,3, combinational circuits (CC1, CC2); 4, 5, blocks of m AND gates; 6, block of m OR gates; 7, random bit generator

The resulting set of logic equations allows one to construct the functional diagram of the generator for the m-bit patterns of the de Bruijn sequence. By implementing the system and adding the equiprobable binary number generator to the device, we obtain a combined sequence generator, whose block diagram is shown in Fig. 6.2. Here block 1 represents a static register, which is fed with a successive pattern on arrival of a clock pulse to its control input. Combinational circuits CC1 (block 2) and CC2 (block 3) are designed in accordance with the set of equations (6.19) and generate m-bit segments of a de Bruijn sequence $A(k + 1) = a_1(k + m)a_2(k + m)\cdots a_m(k + m)$ and $A(k + 2) = a_1(k + 2m)a_2(k + 2m)\cdots a_m(k + 2m)$. Blocks 4 to 6 are standard computing elements consisting of m AND gates, m AND gates with inverting inputs, and m OR gates, respectively. Block 7 is a generator of the random binary digit sequence $b(k) \in \{0, 1\}$, $k = 0, 1, 2, \ldots$, for which the $P[b(k) = 1] = 0.5 - \xi$ condition is met, with $0 < \xi < 1$.

Depending on the value of $b(k)$ in the kth cycle of the composite sequence generator (see Fig. 6.2), an m-bit travels via m AND gates of block 4 or 5 and then via m OR gates of block 6 to the inputs of register 1 where, when written, it becomes the initial segment $A(k)$ for a successive generator cycle. The timing chart of such a device is given in Fig. 6.3 for a generating polynomial $\varphi(x) = 1 \oplus x^3 \oplus x^4$.

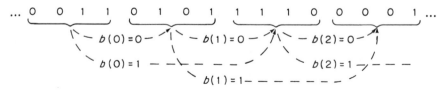

Fig. 6.3. Timing chart of a composite sequence generator for a generating polynomial $\varphi(x) = 1 \oplus x^3 \oplus x^4$

The properties of a composite sequence generated by the device of Fig. 6.2 are discussed in [47]. The properties of such sequences maximally approximate to those of a random number sequence, thereby allowing them, when used as test patterns, to meet conditions $p_1 = 0$ and $P_n = 0$ as $N \to \infty$; the latter is a measure of test sequence quality.

A more significant measure of test sequence quality is how fast $P_n(N)$ converges to zero as N increases, which defines the test sequence length. Therefore real automated test systems, when generating pseudorandom test patterns, solve the problem of speeding up $P_n(N)$ convergence to zero, and hence affect the test sequence length.

Consider now some methods that allow one to improve the quality of test sequences by generating the required probabilities for a 1 to appear in any bits of their test patterns.

6.4. DESIGN OF SEQUENCES WITH SPECIFIED PROBABILITY AND TIME RESPONSES

The application of PRTP generators in digital circuit testing predefines straightforward digital circuit analysis and offers the strong possibility of revealing abnormal sensitivity to unpredictable sequences of input signals [98]. As already mentioned, the method is not always efficient in test time utilization since a proportion of the test patterns applied to the DC do not account for more faults than the test patterns used earlier. Also, bad (invalid) test patterns can occur.

The first disadvantage mentioned can be eliminated by allowing the PRTP generator to assume the probability of 1s or 0s on each of its outputs. Therefore the number of useless input patterns—those whose elimination does not affect the DC test coverage—is reduced and the convergence of $P_n(N)$ to p_1 speeds up.

The second disadvantage—PRTP invalidity—can be eliminated by individually setting the pattern generator to a required set of probability and time responses.

An original PRTP generator that has been discussed in [98] and used in the automated test and diagnostic system CODIAC permits the generation of valid pseudorandom test pattern sequences of reasonable length. However, the speed of PRTP generation is not high owing to the use of a sequential procedure for pattern generation, with variable probabilities of 1s to occur in their positions. A simplified block diagram demonstrating generation of successive positions of the ith generator bit can be represented as a normal D-type flip-flop whose D input is successively fed with independent random bits $a_j \in \{0, 1\}$, $j = 1$ to l, whose probability $P(a_j = 1)$ of 1 to occur is p. The number l of random bits a_j is defined by the number of control signals applied to input C of the D flip-flop. As a result, a binary variable z_i whose probability $P(z_i = 1)$ is defined by the expression

$$P(z_i = 1) = 1 - (1 - p)^l$$

is generated on the direct output of the D flip-flop. The probability $P(\bar{z}_i = 1)$ of a 1 occurring on the inverted output is calculated as

$$P(\bar{z}_i = 1) = 1 - P(z_i = 1) = (1 - p)^l.$$

From the equations obtained it is evident that a wide variation range for probabilities $P(z_i = 1)$ and $P(\bar{z}_i = 1)$ is ensured by significantly increasing the value l or, what amounts to the same thing, decreasing PRTP generation rate.

A classical scheme for a single-cycle generator producing a sequence of values z_i with the specified probability $P(z_i = 1)$ [72] consists of a random number generator, a decoder and a multiple-input OR gate. The number of decoder outputs g connected to the OR gate inputs determines the value of probability $P(z_i = 1)$:

$$P(z_i = 1) = \frac{g}{2^l} \qquad g \in \{0, 1, 2, \dots, 2^l\}. \qquad (6.20)$$

Equation (6.20) points to the possibility of generating a bit sequence z_i with probability $P(z_i = 1)$ varying from 0 to 1 and an increment of $1/2^l$. However, the implementation complexity of such a structure, and its decoder in particular, makes the technique impracticable for PRTP generators.

A more adequate scheme for generating the values of z_i is the one consisting of a pseudorandom number generator and a probability and time response converter for test patterns [47]. This structure ensures individual setting of PRTP generator bits and expands the class of generated test patterns.

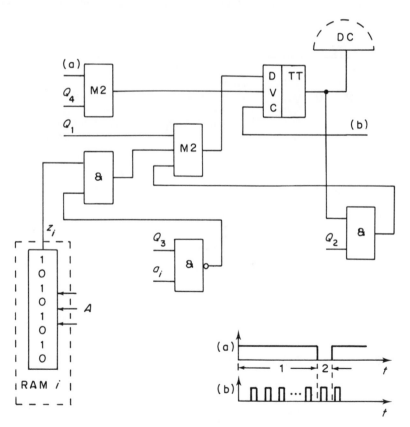

Fig. 6.4. A functional scheme and timing chart for the ith stage of the probability and time response converter: Q_1, Q_2, Q_3, Q_4, controls; A, address inputs; a_i, equiprobable bit values fed from pseudorandom sequence generator

The pseudorandom number generator used here is the one generating uniformly distributed pseudorandom numbers with the probability of a 1 at every output of the generator being 0.5. The design requirements for such devices are discussed in Sec. 6.3.

The functional scheme of a single probability and time response converter stage (Fig. 6.4) consists of an output flip-flop, two two-input AND gates, a two-input NAND gate, two-input and three-input modulo-2 adders and a random access memory of b bits capacity.

Depending on the values of control signals Q_1, Q_2, Q_3 and Q_4, sequences obtained at the ith stage output in the converter have different time and probability responses. The converter function is realized in a cycle consisting of two phases (Fig. 6.4). In the first phase, clock signals applied to the clock inputs of the digital circuit under test are processed. In the second phase, the address inputs of the RAM are successively fed with uniformly distributed pseudorandom numbers, and in the first phase, addresses A that have been produced as counter sequences are fed there.

Table 6.4 gives the values of control signals Q and the contents of the RAM for all possible modes of converter operation by the ith input; X means that a variable may take on the value of 0 or 1. In mode 1, the converter generates by the ith output

Table 6.4. Modes of converter stage operation.

Mode no.	Q_1	Q_2	Q_3	Q_4	Contents of RMA b bits in size	Remarks
1	0	0	0	0	Synchronization sequence	Generation of synchronization sequence
2	0	0	0	1	$111\cdots1$	Const.1
3	1	0	0	1	$111\cdots1$	Const.0
4	0	0	X	1	$000\cdots0$	Const.0
5	1	0	X	1	$000\cdots0$	Const.1
6	X	0	1	1	$111\cdots1$	$P(y_i = 1) = 0.5$
7	0	0	1	1	$\underbrace{111\cdots1}_{g}\underbrace{000\cdots0}_{b}$	$P(y_i = 1) = g/2b \leqslant 0.5$
8	1	0	1	1	$\underbrace{111\cdots1}_{g}\underbrace{000\cdots0}_{b}$	$P(y_i = 1) = 1 - g/2b$, $g/2b \leqslant 0.5$
9	1	1	X	1	$000\cdots0$	Meander-type pulse sequence
10	X	1	1	1	$111\cdots1$	$P(f) = 0.5$
11	0	1	1	1	$\underbrace{111\cdots1}_{g}\underbrace{000\cdots0}_{b}$	$P(f) = g/2b \leqslant 0.5$
12	1	1	1	1	$\underbrace{111\cdots1}_{g}\underbrace{000\cdots0}_{b}$	$P(f) = 1 - g/2b \leqslant 0.5$

a synchronization sequence that has been written as a combination of 0s and 1s in the RAM associated with the stage. The timing chart of the synchronization sequence is produced on the output flip-flop within the first phase by reading data from the RAM. In mode 2, a logic 1 generated at the ith output remains unchanged in the course of converter operation. The contents of the output flip-flop are maintained constant by providing a 1 at the RAM output for every address A that is a random number. Similarly, signals const.0 and const.1 are produced in modes 3 to 5. In the specified modes, the RAM is preloaded with 1s or 0s in all locations (Table 6.4). When $g \in \{1, 2, 3, \ldots, b\}$ locations of the RAM are loaded with 1s and $(b - g)$ locations with 0s, the probability of obtaining a 1 at its output after reading data from a randomly selected location (by the uniformity law) will be $P(z_i = 1) = g/b$. For the values $Q_1 = Q_2 = 0$, $Q_3 = 1$, a sequence of random bits will be applied to the flip-flop input with the probability of a 1 occurring being defined by the expression $P(z_i = 1)P(\bar{a}_i = 1)$, and thus a $y_i = 1$ value will be formed on the ith output of the converter with probability [72]:

$$P(y_i = 1) = P(z_i = 1)P(\bar{a}_i = 1).$$

Modification of the values $P(z_i = 1)$ yields a random sequence of $y_i = 1$ values at the converter output whose probability can be specified in the range $1/2$ to $1/2b$ in increments of $1/2b$. The values of variables Q_n are given in Table 6.4 (modes 6 and 7).

For $Q_1 = 1$, the probability of a $y_i = 1$ value occurring has the form [72]

$$P(y_i = 1) = 1 - P(z_i = 1)P(\bar{a}_i = 1)$$

thereby allowing us to obtain the sequence y_i with a probability of a 1 at the ith output of the converter closely approaching unity.

Modes 9 to 12 are realized on the ith converter output by feeding back the flip-flop output to its input via a modulo-2 adder. In this case the output flip-flop functions in the complementing mode and performs modulo-2 addition on both stored data and those fed to its input in each cycle. In mode 9, meander-type sequence is generated on the flip-flop, i.e. in each cycle the value of y_i is inverted. At the same time, in modes 10 to 12 the value of y_i is inverted with probability $P(f)$ determined by the contents of the RAM. Therefore in modes 10 to 12 the inversion probability is specified for y_i. With $P(y_i = 1) = 0.5$ the probability variation increment $P(f)$ is $1/2b$.

The above operating modes of a PRTP generator consisting of a pseudorandom or random number generator and a converter (Fig. 6.4) can be realized simultaneously for different positions of device stages.

To ensure correct time interrelations between any converter positions, provision is made in the converter for bit-group modes, with the number of bits in the group as well as their total number not exceeding the RAM size. A group of $v \leqslant b$ outputs is formed of any converter positions by writing data to the RAMs associated with the selected positions according to Table 6.5, where the rows contain position numbers and the columns contain the numbers of RAM locations forming a group.

The values of variables Q_1, Q_2, Q_3 and Q_4 specify the kind of group modes that are listed in Table 6.6.

In modes 1 and 2, the test pattern sequences produced are of the 'walking 1' or 'walking 0' type, whereby only one bit of v can be 1 or 0, respectively. In modes 3 and 4, only one bit of v will be 1 or 0 with probability 0.5. In modes 1 to 4, the

Table 6.5.

No.	1	2	3	...	v
1	1	0	0	...	0
2	0	1	0	...	0
3	0	0	1	...	0
⋮	⋮	⋮	⋮		⋮
b	0	0	0	...	1

Table 6.6.

Mode	Q_1	Q_2	Q_3	Q_4	Pattern type
1	0	0	0	1	Walking 1
2	1	0	0	1	Walking 0
3	0	0	1	1	Walking 1, $P(1) = 0.5$
4	1	0	1	1	Walking 0, $P(0) = 0.5$
5	0	1	0	1	Gray code
6	1	1	0	1	Inverse Gray code
7	0	1	1	1	Pseudocyclic code
8	1	1	1	1	Inverse pseudocyclic code

number of the bit whose value is 0 or 1 is determined by a successive random number code applied simultaneously to the address inputs of RAMs for all converter positions.

The Gray code generated in mode 5 is a PRTP sequence whose adjacent test patterns differ by only one bit from one another. The inverse Gray code is characterized by the fact that adjacent test patterns agree in only one bit (mode 6).

The pseudocyclic codes generated in mode 7 are taken to mean pseudorandom numbers with the possibility of modifying only one bit at a time in a test pattern that has been produced in the preceding cycle with probability 0.5. At the same time, inverse pseudocyclic codes are those ensuring the probability 0.5 for only one bit to remain unmodified in the output test pattern.

The group modes of a PRTP generator consisting of a pseudorandom number generator and a converter can be implemented simultaneously for different groups of generator bits with their number and sizes being defined only by b.

As an example of a PRTP generator, consider a 6-bit generator that has been designed for testing combinational circuits (Fig. 6.5). Its function is controlled by clock pulses applied to the C inputs of output flip-flops. The V inputs of the flip-flops are fed with logic 1s since the working cycle of a PRTP generator used for combinational circuit testing consists of only one phase (the second). Fig. 6.5 shows the equivalent scheme of a PRTP generator for the case when signal const.0 is produced by its first bit, the meander-type signal is produced by its second bit, and a pseudorandom sequence y_4 with probability $P(y_4 = 1) = \frac{1}{8}$ is produced by its fourth bit. The third, fifth and six bits form a group at the outputs of which pseudorandom patterns of 'walking 1' type are produced. The values of variables Q_1, Q_2, Q_3 and Q_4 for the first, second and fourth bits of the generator (Fig. 6.5) are respectively

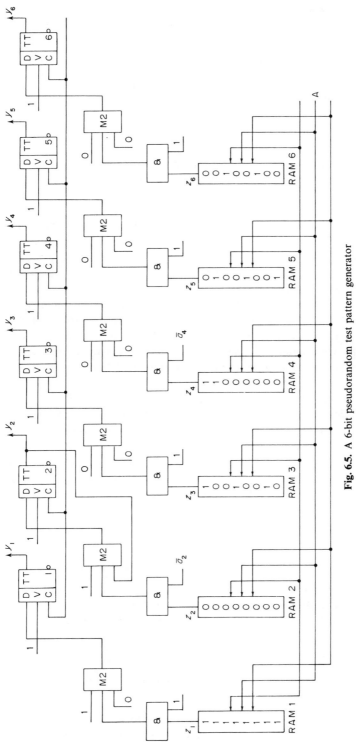

Fig. 6.5. A 6-bit pseudorandom test pattern generator

Table 6.7.

A	y_1	y_2	y_3	y_4	y_5	y_6
001	0	1	0	1	1	0
100	0	0	0	0	1	0
010	0	1	0	0	0	1
101	0	0	0	0	0	1
110	0	1	1	0	0	0
111	0	0	0	0	1	0
000	0	1	1	1	0	0
100	0	0	0	0	1	0
010	0	1	0	0	0	1

specified by modes 3, 9 and 7 (see Table 6.4), and for the remaining bits by mode 1 (Table 6.6). The RAM associated with every bit consists of eight locations.

The PRTP sequence, each pattern of which consists of $y_1 y_2 y_3 y_4 y_5 y_6$ as a function of the sequence A of addresses represented by uniformly distributed pseudorandom numbers, is given in Table 6.7.

The analysis of sequences generated in accordance with the table demonstrates the extensive functional possibilities of the PRTP generator implemented with the use of probability and time response converters of test patterns. This allows us to decrease significantly the number of useless test patterns by setting each of its bits to the desired probability and time response. Thereby $P_n(N)$ converges to zero faster and the test sequence length is diminished.

The probability response converter of Fig. 6.4 is based on the principle of varying the probability for a 1 to appear on its specified outputs and specifying the required cross-correlation function for their sequences. This approach can be further developed by using dependent sequences of pseudorandom patterns [72, 110]. Let us demonstrate this by the portion of a digital circuit of Fig. 6.6. It consists of a shift register and an R–S flip-flop for which input variables x_1 and x_2 must satisfy the condition $x_1 x_2 \neq 1$. Thus preliminary examination of the digital circuit shows that for bits $y_1(k-1)$ and $y(k)$ of a pseudorandom test sequence $\{y_1(k)\}$, $k = 1$ to l, the equation

$$y_1(k-1)y_1(k) \neq 1 \qquad (6.21)$$

must be satisfied.

A device converting the sequence $\{a_1(k)\}$, $k = 1$ to l, of equiprobable bits into the sequence $\{y_1(k)\}$ for which the relation (6.21) holds can be implemented by a technique of [110]. Such a converter consists of a delay element and a two-input AND gate (Fig. 6.6). Any successive bit of $y_1(k) = a_1(k)\overline{y_1(k-1)}$ and the product $y_1(k-1)y_1(k)$ can be calculated by

$$y_1(k-1)y_1(k) = y_1(k-1)a_1(k)\overline{y_1(k-1)} = 0$$

and meets the condition (6.21).

Thus the use of pseudorandom test sequences for digital circuit testing allows one to solve the problem basically as well as to achieve reasonable execution time.

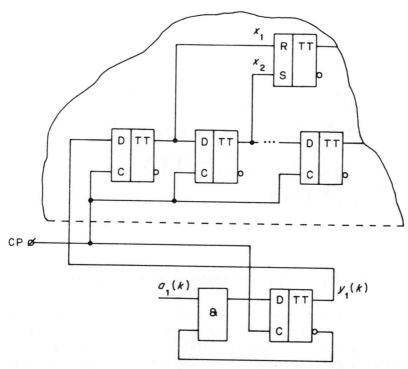

Fig. 6.6. An example of generating dependent sequences of pseudorandom signals $y_1(k)$ for testing a sequential circuit

However, for the technique to be applied for testing sequential circuits of arbitrary configuration, then they must be thoroughly analysed in advance, which is in many cases rather laborious. Therefore, the use of pseudorandom sequences for testing LSI and VLSI circuits requires special methods and techniques that allow one to decrease and sometimes to eliminate the laborious analysis of their topology.

The basic idea behind the use of pseudorandom sequences as tests for diagnosing LSI and VLSI faults is their implementation by probabilistic or trivial testing techniques. In spite of the use of long test sequences, probability testing in practice does not necessarily ensure high-quality testing for LSI and VLSI circuits. Diagnostics reliability is improved by implementing a trivial testing technique (exhaustive testing) that consists of generating all possible input patterns for individual portions of a digital circuit. Next we shall discuss the peculiarities of this technique as applied to LSI and VLSI testing.

7

APPLICATION OF PSEUDORANDOM SEQUENCES TO VLSI TESTING

7.1. EXHAUSTIVE TESTING OF VLSI CIRCUITS

This chapter deals with an efficient solution to the problem of built-in hardware design for VLSI self-testing. The case in point is exhaustive testing of the combinational part of a VLSI chip [111–114]. This approach implies the use of the LSSD technique in the test mode. The idea behind it is that the VLSI structure is represented as a shift register and a combinational portion partitioned into multiple combinational subcircuits. Partitioning of the VLSI structure into two parts is commonly used in the self-test design [40].

Exhaustive testing of VLSI circuits is a procedure based on generation of all possible input patterns for each combinational subcircuit in the VLSI circuit.

Consider in more detail the VLSI structure in the test mode. In this case, the scheme will consist of shift register P1 of length w and combinational circuit G consisting or r subcircuits G_i, $i = 1$ to r, with the ith subcircuit having n_i inputs and v_i outputs (Fig. 7.1). Let G have n inputs and v outputs. Normally $v_i = 1$, hence $v = r$.

It should be noted that, in the general case, the self-testing VLSI structure consists of a more complex combination of shift registers and combinational circuits. However, the classic problem of generating an exhaustive test sequence for each subcircuit in the combinational portion may be solved for any such modification. For the VLSI circuit of Fig. 7.1, the problem consists of generating a test sequence $\{y(k)\}$, $k = 1$ to l, that satisfies the following conditions.

(i) The sequence of symbols $y(k) \in \{0, 1\}$, $k = 1$ to l, applied to an input of shift register P1 of length w ensures generation of all possible binary patterns for any subcircuit G_i, $i = 1$ to r, in G.

(ii) The generated sequence length l takes on the minimal possible value, which is normally less than 2^w, thereby allowing one to make the testing procedure on-line.

The problem of $\{y(k)\}$ generation can be solved by different methods and approaches [45]. However, the major measure of their effectiveness is the complexity

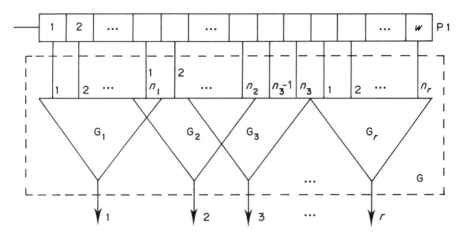

Fig. 7.1. Testing of a VLSI circuit in an LSSD environment

of constructing the $\{y(k)\}$ generator, which sometimes is higher than that of VLSI chip implementation.

One of the most efficient solutions available is the use of an M-sequence as $\{y(k)\}$. Then hardware overheads will be m bits of P1, where $m = \deg \varphi(x)$ is the degree of generating polynomial $\varphi(x)$, which meets the condition

$$\deg \varphi(x) > w \qquad (7.1)$$

and some modulo-2 adders. The length of test sequence N is defined by

$$N = 2^m - 1 > 2^w.$$

However, for actual values $w = 100\text{--}300$ [114], the test sequence length N becomes prohibitive for on-line exhaustive testing. Expression (7.1) yields only an upper bound on the degree of generating polynomial $\varphi(x)$ that allows one to implement exhaustive testing. The lower bound can be calculated from the inequality

$$\deg \varphi(x) > \max_i n_i. \qquad (7.2)$$

Note that the inequality (7.2) gives a necessary but not sufficient condition to be satisfied by the generating polynomial $\varphi(x)$. In fact, for the combinational portion of circuit G in Fig. 7.2, which consists of a single subcircuit G_1 with $n_1 = 3$ inputs, it is impossible to ensure all possible input patterns by using $\varphi(x) = 1 \oplus x^3 \oplus x^4$ with $\deg \varphi(x) = 4 > n_1 = 3$. Then there will be no input patterns 001, 010, 100 on inputs 1, 2 and 3 of G, which is defined by the form of the selected polynomial $\varphi(x)$. Based on the shift-and-add property of the M-sequence generated by $\varphi(x) = 1 \oplus x^3 \oplus x^4$, we can demonstrate that the relation $a_1(k) = a_4(k) \oplus a_5(k)$, $k = 1$ to l, is met for values $a_1(k)$, $a_4(k)$ and $a_5(k)$ of shift-register positions 1, 4 and 5. As a consequence, some input patterns are missing and, hence, certain faults such as $\equiv 0$ are undetectable at the output of a two-input AND gate.

Thus it is necessary to select a suitable polynomial $\varphi(x)$ to generate the sequence $\{a(k)\}$, $k = 1$ to $2^m - 1$, which meets the conditions below.

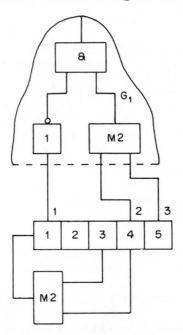

Fig. 7.2. Testing a digital circuit by pseudorandom patterns produced by the generator described by
$\varphi(x) = 1 \oplus x^3 \oplus x^4$

(i) On generating an M-sequence $\{a(k)\}$, $k = 1$ to $2^m - 1$, described by $\varphi(x)$, all possible n_i-bit patterns are produced in the positions of shift register P1 (see Fig. 7.1) whose numbers Q_c^i, $c = 1$ to n_i, are determined by the topology of connections between the inputs of the ith subcircuit G_i and register P1. This condition must be satisfied for each subcircuit G_i.

(ii) The high degree m of generating polynomial $\varphi(x)$ assumes the minimal possible value defined by the inequality

$$\max_i n_i < m < w + 1$$

or, subject to generation of a zero n_i-bit code, where $i = 1$ to r, as a result of initial setting of P1, by the inequality

$$\max_i n_i \leqslant m \leqslant w. \tag{7.3}$$

To ensure the above conditions for the generating polynomial $\varphi(x)$, the circuit under test should be subjected to a rigorous analysis. Later on we shall discuss other approaches to performing the analysis as well as estimating the efficiency of their application.

This analysis can be eliminated from the procedure of synthesizing the required sequence $\{a(k)\}$ through the definition of its generating polynomial only by increasing the period and respectively the length N of the test sequence. Here we can confine ourselves to the relation (7.1) or expression

$$\deg \varphi(x) = \max_i \left(\max_c Q_c^i - \min_c Q_c^i \right) + 1 \tag{7.4}$$

where Q_c^i is the number of positions of P1 (see Fig. 7.1) whose output is connected to the cth input, $c = 1$ to n_i, of the ith subcircuit G_i.

For the example shown in Fig. 7.2, the set of combinational subcircuits G_i consists of a single subcircuit G_1 for which $c = 1$ to 3 and $Q_1^1 = 1$, $Q_2^1 = 4$ and $Q_3^1 = 5$.

From equation (7.4), we obtain $\deg \varphi(x) = Q_3^1 - Q_1^1 + 1 = 5 - 1 + 1 = 5$; hence, the application of $\varphi(x) = 1 \oplus x^2 \oplus x^5$, for examples, ensures exhaustive testing of the circuit G_1 under test (Fig. 7.2). The test sequence length N will be equal to $2^5 - 1 = 31$.

Thus by equation (7.4), we can calculate the high degree of polynomial $\varphi(x)$ that ensures exhaustive testing of the circuit under test without its analysis. Then any primitive polynomial $\varphi(x)$ whose degree is defined by (7.4) will meet the requirements imposed on such polynomials.

Returning to the example of interest, we may easily demonstrate that the combinational circuit G_1 may also be exhaustively tested by application of polynomial $\varphi(x) = 1 \oplus x^2 \oplus x^3$. In fact, all possible 3-bit combinations are applied to the three input lines to combinational circuit G_1 in accordance with the time chart for the states of memory elements in the shift register (Fig. 7.3). This time the length of test sequence $N = 9$ will be significantly shorter than that for the case of polynomial $\varphi(x) = 1 \oplus x^2 \oplus x^5$.

As is seen, an attempt to eliminate the procedure of synthesizing an optimal M-sequence in terms of its minimally possible period causes the test sequence length N to be significantly increased. In this case, the VLSI self-test is implemented by any M-sequence whose high degree satisfies the relation (7.4). At the same time, by artificially synthesizing the M-sequence, we can find the polynomial $\varphi(x)$ whose high degree is defined by $m = \max_i n_i$ [114]. Then the minimal possible length N of test sequence $\{a(k)\}$, which is an M-sequence, can be found by

$$N = 2^m + w - m - 1 \tag{7.5}$$

where w is the number of positions in shift register P1 (see Fig. 7.1). The term $w - m$ is defined by the requirement of producing all possible m-bit combinations for the specified topology of connections between G_i and P1 with a random shift relative to the register positions; $w - m$ and $2^m - 1$ are respectively the lengths of the homing and testing steps whose sum is the length N of test sequence $\{a(k)\}$. However, the value of N does not account for a zero m-bit combination produced by initializing P1 to zero.

For the practical case, when at least one of the P1 positions (see Fig. 7.1) fails to be initialized to the specified states, an alternative solution based on the use of de Bruijn sequences has been proposed [47]. The latter allows one to generate the complete set of 2^m binary combinations in $m = \max_i n_i$ sequential positions of shift

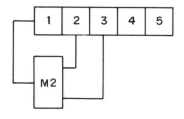

Fig. 7.3. Generator of an M-sequence described by the polynomial $\varphi(x) = 1 \oplus x^2 \oplus x^3$

Table 7.1.

Ref.	Position no. in shift register					Circuit input no.		
	1	2	3	4	5	1	2	3
1	1	0	0	0	0	1	0	0
2	0	1	0	0	0	0	0	0
3	1	0	1	0	0	1	0	0
4	1	1	0	1	0	1	1	0
5	1	1	1	0	1	1	0	1
6	0	1	1	1	0	0	1	0
7	0	0	1	1	1	0	1	1
8	1	0	0	1	1	1	1	1
9	0	1	0	0	1	0	0	1

register P1. However, an implementation of such a sequence generator requires a supplementary $(m-1)$-input AND gate and a two-input EOR gate to be inserted. Furthermore, the minimal length of the test sequence increases by 1.

For the example of Fig. 7.3, all possible binary combinations are produced in any set of sequential positions of the shift register of length $\max_i n_i = 3$. An exception is the combination 000, which is not available in the first three register positions (Table 7.1). This may be attributed to the fact that the test sequence used is an M-sequence described by polynomial $\varphi(x) = 1 \oplus x^2 \oplus x^3$. By using the given polynomial, we can implement the de Bruijn sequence generator defined by the set of logic equations [47]

$$a_1(k+1) = a_2(k) \oplus a_3(k) \oplus [\overline{a_1(k)}\,\overline{a_2(k)}]$$
$$a_2(k+1) = a_1(k) \qquad a_3(k+1) = a_2(k).$$

This generator allows one to produce all possible binary combinations in any three sequential positions of shift register P1. As seen from Fig. 7.4, the implementation of such a generator is more expensive from the hardware point of view.

Thus, the basic problem that arises on implementing the VLSI self-test by the use

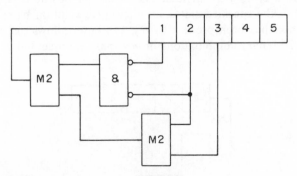

Fig. 7.4. An implementation of a de Bruijn sequence generator

of M-sequences is to select a polynomial $\varphi(x)$ for generating a sequence $\{a(k)\}$ which ensures exhaustive testing of the combinational portion. We shall consider several solutions to the problem dependent on the requirements of the VLSI chip.

7.2. DESIGN OF TEST SEQUENCE GENERATORS

Let us discuss some possible approaches to solving the problem of designing test sequence generators used for implementing the VLSI self-test (see Fig. 7.1). The structure presented implies that the topology of connections between each combinational subcircuit G_i, $i = 1$ to r, of circit G and the shift register P1 can assume an arbitrary form. Then the unique attribute denoting the interconnection between G_i and the shift-register positions is the set of their numbers $Q_1^i, Q_2^i, Q_3^i, \ldots, Q_{n_i}^i$, where n_i is the total number of G_i inputs. On the grounds of the given values Q_c^i, $c = 1$ to n_i, we can calculate

$$Q^i = \max_c Q_c^i - \min_c Q_c^i \tag{7.6}$$

which shows the layout of connections in the circuit under test. Thus, for $Q^i = n_i - 1$, all G_i inputs are connected to the sequential positions of the shift register, and for $Q^i = w - 1$ they are connected to the designated n_i positions lying between the first and the last positions of P1. The first case is then more suitable for designing the generator of a test sequence, which is a pseudorandom M-sequence. In fact, for

$$Q = \max_i Q^i \tag{7.7}$$

which takes on small values (20–30), the problem of designing a pseudorandom test sequence generator reduces to a trivial selection of polynomial $\varphi(x)$ whose high degree corresponds to (7.4) and equals $Q + 1$. For higher values of Q, we are faced with the problem of designing the generator to provide all possible $2m$ binary combinations on any $m = \max n_i$ positions of the $w > m$ positions of the shift register P1.

Let us solve the problem of test sequence generator design for the example of Fig. 7.2. First of all, we shall prove the following theorem [115].

Theorem 7.1. The set of expressions of the form $a(k) = a(k + p_1) \oplus a(k + p_2)$ that hold true for the symbols of an M-sequence described by the primitive polynomial $\varphi(x)$ does not intersect the set of expressions $b(x) = b(k + g_1) \oplus b(k + g_2)$ that hold true only for the symbols of an M-sequence described by polynomial $\varphi^{-1}(x)$ conjugate to $\varphi(x)$ if and only if $L \bmod 3 \neq 0$, where $L = 2^m - 1$, $m = \deg \varphi(x)$, $k = 0$ to $L - 1$.

Proof. Let a primitive polynomial $\varphi(x)$ generate an M-sequence $\{a(k)\}$, $k = 0$ to $L - 1$, and its conjugate polynomial $\varphi^{-1}(x)$ generate a sequence $\{b(k)\}$. Assuming that $\varphi^{-1}(x)$ produces an inverse sequence, we obtain

$$\{a(k)\} = \{b(L - k - 1)\} \qquad k = 0 \text{ to } L - 1. \tag{7.8}$$

By the shift-and-add property of the M-sequence $\{a(k)\}$, we may write the relation

$$a(k) = a(k + p_1) \oplus a(k + p_2) \qquad p_1 > p_2 \tag{7.9}$$

which implies that there is dependence between its symbols $a(k)$, $a(k + p_1)$ and $a(k + p_2)$ for specific p_1 and p_2.

Using expression (7.8), we rearrange equation (7.9) as

$$b(L - k - 1) = b(L - k - p_1 - 1) \oplus (L - k - p_2 - 1). \qquad (7.10)$$

At the same time, for the sequence symbols $\{b(k)\}$ an equation similar to equality (7.9) is true:

$$b(k) = b(k + g_1) \oplus b(k + g_2) \qquad g_1 > g_2. \qquad (7.11)$$

Expressions (7.10) and (7.11) hold true for all k and their simultaneous solution for $g_1 = p_1$ and $g_2 = p_2$ proves that the set of equations (7.9) intersects the set of similar identities (7.11).

We next define the conditions for simultaneous solution of equations (7.9) and (7.11). For that purpose, we obtain the system of two identities

$$b(p_1) = b(0) \oplus b(p_1 - p_2)$$
$$b(0) = b(p_1) \oplus b(p_2)$$

which is only solvable for $p_2 = p_1 - p_2$ (i.e. $p_1 = 2p_2$) on the basis of equations (7.10) (with $k = L - p_1 - 1$) and (7.11) (with $k = 0$ and $g_1 = p_1$, $g_2 = p_2$).

With $k = L - p_2 - 1$ from (7.10) and $k = 0$ from (7.11), we find

$$b(p_2) = b(L + p_2 - p_1) \oplus b(0)$$
$$b(0) = b(p_1) \oplus b(p_2)$$

for $g_1 = p_1$ and $g_2 = p_2$ [47]. This system holds true if $p_1 = L + p_2 - p_1$, i.e. $2p_1 = L + p_2$.

Thus simultaneous solution of equations (7.9) and (7.11), which show the linear dependence between symbols $a(k)$, $a(k + p_1)$ and $a(k + p_2)$ in the sequence $\{a(k)\}$ and $b(k)$, $b(k + p_1)$ and $b(k + p_2)$ in the sequence $\{b(k)\}$, is determined by equalities $p_1 = 2p_2$ and $2p_1 = L + p_2$, implying that intersection of the sets of (7.9) and (7.11) is indicated by the expression $3p_2 = L$ or, which is the same, by $L \bmod 3 = 0$. Otherwise, with $L \bmod 3 \neq 0$, the sets of specified equations will be disjoint, which proves the theorem. ∎

Corollary. By successively applying M-sequences described by $\varphi(x)$ and $\varphi^{-1}(x)$, we may obtain test sequences that produce all possible 3-bit combinations in any three positions of shift register P1 (see Fig. 7.1). It should be noted, however, that a zero combination is ensured in the case of initializing the shift register P1.

Thus the use of primitive polynomials $\varphi(x)$ and $\varphi^{-1}(x)$, for which $L \bmod 3 \neq 0$, ensures that for any subcircuit G_1 of G all possible input patterns are produced at the $n_i \leqslant 3$ inputs connected to n_i arbitrary input register positions.

For the example of Fig. 7.2, the number of inputs n to the only subcircuit G_1 of G is 3 and $Q^1 = \max_c Q_c^1 - \min_c Q_c^1, c = 1$ to 3, the same as $Q = \max_i Q^i = 4$. Hence, there may be two alternative approaches to self-test design for the circuit under test.

The first one is based on the use of a primitive polynomial $\varphi(x)$ of deg $\varphi(x) = Q + 1 = 5$. As has been shown earlier, the use of polynomial $\varphi(x) = 1 \oplus x^2 \oplus x^5$, for example, allows one to generate all possible 3-bit combinations on the inputs to circuit G_1 (see Fig. 7.2). However, the test sequence length $N = 31$ will then significantly exceed the lower bound $N = 9$ determined by equation (7.5).

The second alternative is based on the corollary to Theorem 7.1, according to which the use of polynomials $\varphi(x) = 1 \oplus x^2 \oplus x^3$ and $\varphi^{-1}(x) = 1 \oplus x \oplus x^3$ allows one to obtain all possible binary combinations on three inputs to circuit G_1 (see Fig. 7.2).

A similar result involving the generation of all possible combinations of $v \leqslant 3$ bits on v inputs to circuit G_i may be obtained by sequentially applying the direct $a(k)$ and inverse $\overline{a(k)}$ values of M-sequence symbols. This may be proved by the following theorem.

Theorem 7.2. The sequential generation of direct $a(k)$ and inverse $\overline{a(k)}$ values of symbols of an M-sequence described by polynomial $\varphi(x)$ allows one to produce all possible binary combinations in any $v \leqslant 3$ positions of the shift register whose length w satisfies the inequality $w \leqslant 2^m - 1$, where $m = \deg \varphi(x)$, $k = 1$ to $2^m - 1$.

Proof. By generating direct values $a(k)$ of symbols for the M-sequence $\{a(k)\}$, where $k = 1$ to $2^m - 1$, we can obtain all possible combinations of $v \leqslant 2$ bits. This stems from the properties of sequences generated by a primitive polynomial [85]. All combinations of $v \leqslant 2$ bits will be produced in any v positions of a shift register of length $w \leqslant 2^m - 1$. However, all possible combinations can be obtained for the selected $v = 3$ positions of the specified register if and only if there is no linear dependence between symbols $a(k)$, $a(k + p_1)$ and $a(k + p_2)$ produced on them. Otherwise, for the symbols linearly related by

$$a(k) \oplus a(k + p_1) = a(k + p_2) \tag{7.12}$$

only binary combinations of the form 000, 101, 011 and 110 are produced in the appropriate shift-register positions but combinations 001, 100, 010 and 111 are missing.

For the inverse values $\overline{a(k)}$ of sequence $\{a(k)\}$, the inverse codes $\overline{a(k)}$, $\overline{a(k + p_1)}\ \overline{a(k + p_2)}$ are generated in the shift-register positions determined by (7.9). Then code $a(k)a(k + p_1)a(k + p_2) = 000$ is associated with $\overline{a(k)}\ \overline{a(k + p_1)}\ \overline{a(k + p_2)} = 111$, code 010 is associated with 010, 011 with 100 and 110 with 001. Thus by using the direct $a(k)$ and inverse $\overline{a(k)}$ values of the symbols of the M-sequence $\{a(k)\}$, all possible binary combinations are generated on any $v \leqslant 3$ shift-register positions whose number satisfies the condition $w \leqslant 2^m - 1$. This proves the theorem. ∎

An implementation of a generator producing inverse values $\overline{a(k)}$ of sequence $\{a(k)\}$ differs from the classical structure of an M-sequence generator in an extra input to the moduo-2 adder that has been inserted in the shift-register feedback loop. For the M-sequence generator described by polynomial $\varphi(x) = 1 \oplus x^2 \oplus x^3$, a functional scheme of the device generating the inverse sequence $\overline{a(k)}$ is given in Fig. 7.5. The device will generate the specified sequence with a logic 1 present at the extra input to the modulo-2 adder. For generation of an $\{a(k)\}$ sequence, a logic 0 is applied to the control input.

The advantage of using direct and inverse values of an M-sequence for producing

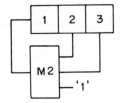

Fig. 7.5. A functional scheme of an inverse M-sequence generator

all possible binary combinations in any $v \leqslant 3$ shift-register positions consists of the possibility of generating a zero 3-bit code which is proved by the timing chart for the states of memory elements in the device (Fig. 7.5).

The two approaches to designing pseudorandom test sequence generators used in the VLSI self-test discussed above are only efficient for VLSI chips consisting of multiple combinational subcircuits G_i, $i = 1$ to r, for which the condition $\max_i n_i \leqslant 3$ is satisfied. For the general case, when $\max_i n_i$ may assume any value, the solution of the problem requires the use of a more fundamental outcome of cyclic code theory and combinatorial analysis. Let us investigate the possibility of applying the postulates of these theories for designing test pattern generators.

It has been shown in [116] how coding theory may be used for implementation of self-testing digital devices. Here the design of an address generator for memory elements which ensures the maximal error detection capability for these elements has been discussed.

The results reported in [111, 113, 117] are based on the use of cyclic codes. A technique of applying such codes is based on the corollary to the following theorem [111], which is given here without proof.

Theorem 7.3. The cyclic code of length l produced by a generating polynomial $\varphi(x)$ and having the minimal code distance d allows one to obtain all possible binary combinations in any $d - 1$ positions of the shift register of length $w \leqslant 1$.

According to Theorem 7.3, the use of M-sequences described by primitive polynomials and hence representing cyclic codes with the code distance $d = 3$ allows one to produce all possible binary combinations in any two positions of the register. Thus by using the polynomial $\varphi(x) = 1 \oplus x \oplus x^3$, which generates a code of length $l = 7$ and has the code distance $d = 3$, we can solve the problem for any two positions in the register of length $w = 7$. Then the test sequence length N will be $2^3 - 1 + 4 = 11$.

To make any three-input subcircuit G_i, $i = 1$ to r, exhaustively testable, a cyclic code with the code distance $d = 4$ is to be used. An example of such a code is that described by polynomial $\varphi(x) = (1 \oplus x)(1 \oplus x \oplus x^3)$ of length $l = 7$ for which the complete set of code words is generated by the scheme of a code generator described by $\varphi(x)$ (Fig. 7.6). Here the positions of generator register must first be set to the initial state described by code 1000 and then the first subset of code words can be generated (Fig. 7.6). Once the shift register has been set to 0111, the second subset of code words (whose total is 7) can be generated. The test sequence length N is 22, the homing steps included. The order of generating the discussed test patterns without homing steps and the complete set of code words ensuring all possible binary combinations in any $v \leqslant 3$ positions of the register of length $w = 7$ are shown in Fig. 7.6.

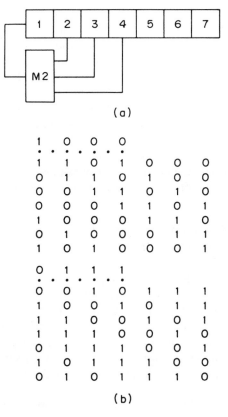

```
1   0   0   0
·   ·   ·   ·   ·   ·
1   1   0   1   0   0   0
0   1   1   0   1   0   0
0   0   1   1   0   1   0
0   0   0   1   1   0   1
1   0   0   0   1   1   0
0   1   0   0   0   1   1
1   0   1   0   0   0   1

0   1   1   1
·   ·   ·   ·   ·   ·
0   0   1   0   1   1   1
1   0   0   1   0   1   1
1   1   0   0   1   0   1
1   1   1   0   0   1   0
0   1   1   1   0   0   1
1   0   1   1   1   0   0
0   1   0   1   1   1   0
```

(b)

Fig. 7.6. (a) A block-diagram of a cyclic code generator with the code distance $d = 4$ and (b) the timing chart of code word generation

A more complex implementation is characterized by application of a cyclic code described by polynomial $\varphi(x) = (1 \oplus x \oplus x^4)(1 \oplus x \oplus x^2 \oplus x^3 \oplus x^4)$ with $l = 15$ and $d = 5$. However, its use does not provide for all possible binary combinations in any $v \leqslant 4$ positions of the shift register with $w \leqslant 15$, and the test sequence length N without homing steps will be calculated as $2^8 - 1$ [111].

The use of the classic Golay code described by polynomial $\varphi(x) = 1 \oplus x \oplus x^5 \oplus x^6 \oplus x^7 \oplus x^9 \oplus x^{11}$ and having length $l = 23$ provides for all possible binary combinations of $v \leqslant 6$ bits in any six out of the 23 positions of the shift register [111].

The above examples of cyclic codes applied to exhaustive testing implementation show that the problem can basically be solved for the specified number of inputs n_i to a combinational subcircuit G_i, $i = 1$ to r, of the combinational VLSI portion whose structure appears as shown in Fig. 7.1 in the test mode. However, the discussed approach capabilities are significantly restricted due to the necessity of presetting the initial codes for generating all subsets of a complete set of code words, thereby requiring that memory elements are included in the VLSI design for storing these codes. An alternative solution based on storing the initial codes supplied from outside the VLSI chip can significantly increase the testing time, which becomes prohibitive for actual values $n_i = 20$ to 30.

An original solution to exhaustive testing design for the VLSI combinational part by implementing the LSSD technique in the test mode has been proposed in [118].

Further development of the discussed method and a comparative estimation of its efficiency are given in [112]. Analysis of the method has disclosed its disadvantages, which are similar to those of the method based on cyclic codes. A major disadvantage of all test sequence generator design methods is the large length of sequences produced by the generators based on them. By using the test sequence generator design methods we can solve the problem of exhaustive testing for n_i-input subcircuit G_i whose inputs are connected to any n_i of w positions in the shift register. In fact, its inputs are connected to specific n_i positions, thereby simplifying the definition of the problem to be solved. Let us discuss the possible ways of exhaustive test design for a specified layout of connections of the VLSI combinational part to the shift-register positions.

7.3. TEST SEQUENCE GENERATOR DESIGN FOR A SPECIFIED VLSI LAYOUT

As was mentioned earlier, the structure of a VLSI chip that implements the LSSD technique in the test mode is uniquely described by the sets of numbers $\{Q_1^i, Q_2^i, \ldots, Q_{n_i}^i\}$ of shift register positions (see Fig. 7.1), which are connected to the inputs of subcircuits $G_i, i = 1$ to r, in the combinational part. The sets of shift-register position numbers give all information required for designing test sequence generators which ensure exhaustive coverage for VLSI chips. The latter is normally based on the use of pseudorandom test sequences, which are generated on the base of primitive generating polynomials.

A major problem to be solved in designing a pseudorandom test sequence generator is to find a generating polynomial $\varphi(x)$ for an M-sequence $\{a(k)\}$, $k = 1$ to $2^m - 1$, which provides for generation of all possible binary combinations on all sets $\{Q_1^i, Q_2^i, \ldots, Q_{n_i}^i\}$ of positions in the shift register of the VLSI chip under test. Consider the solution of the problem for the case when the combinational part of the VLSI chip consists of a single subcircuit G_1 whose layout of input connections is described by the set $\{Q_1, Q_2, \ldots, Q_n\}$.

In the first place, let us associate the set of shift-register position numbers $\{Q_1, Q_2, \ldots, Q_n\}$ with the values of symbols $a(k), a(k + \tau_1), a(k + \tau_2), \ldots, a(k + \tau_{n-1}), k = 0, 1, 2, \ldots$, of the desired M-sequence, where $\tau_j = Q_{j+1} - Q_1$, $j = 1$ to $n - 1$, and $Q_{j+1} > Q_j$. Then the original problem of exhaustively covering the combinational VLSI part G_1 reduces to making the symbols $a(k), a(k + \tau_1), a(k + \tau_2), \ldots, a(k + \tau_{n-1})$ of the desired M-sequence linearly independent. Its generating polynomial $\varphi(x)$ has the high degree m defined by (7.2) whose minimal value is n.

By the shift-and-add property of the M-sequence $\{a(k)\}$, we can show that any symbol $a(k + \tau)$ of the sequence, where $\tau \in \{0, 1, 2, \ldots, 2^m - 2\}$, may be represented as a modulo-2 sum of a subset of symbols $a(k), a(k + 1), \ldots, a(k + m - 1)$ defined by coefficients $\delta_i(\tau) \in \{0, 1\}$, $i = 1$ to m, from the equation [47]

$$a(k + \tau) = \sum_{i=1}^{m} \delta_i(\tau) a(k + i - 1). \tag{7.13}$$

Any τ will be associated with a distinct set of coefficients $\delta_i(\tau)$ [47].

Table 7.2.

τ	$\delta_1(\tau)$	$\delta_2(\tau)$	$\delta_3(\tau)$	$\delta_4(\tau)$	$\delta_5(\tau)$	τ	$\delta_1(\tau)$	$\delta_2(\tau)$	$\delta_3(\tau)$	$\delta_4(\tau)$	$\delta_3(\tau)$
0	1	0	0	0	0	16	1	1	0	1	1
1	0	1	0	0	0	17	1	1	0	0	1
2	0	0	1	0	0	18	1	1	0	0	0
3	0	0	0	1	0	19	0	1	1	0	0
4	0	0	0	0	1	20	0	0	1	1	0
5	1	0	1	0	0	21	0	0	0	1	1
6	0	1	0	1	0	22	1	0	1	0	1
7	0	0	1	0	1	23	1	1	1	1	0
8	1	0	1	1	0	24	0	1	1	1	1
9	0	1	0	1	1	25	1	0	0	1	1
10	1	0	0	0	1	26	1	1	1	0	1
11	1	1	1	0	0	27	1	1	0	1	0
12	0	1	1	1	0	28	0	1	1	0	1
13	0	0	1	1	1	29	1	0	0	1	0
14	1	0	1	1	1	30	0	1	0	0	1
15	1	1	1	1	1						

By way of example illustrating the relation (7.12), we can give the values of coefficients $\delta_i(\tau), \tau \in \{0, 1, 2, \ldots, 30\}$ (Table 7.2), for the symbols of an M-sequence generated by polynomial $\varphi(x) = 1 \oplus x^3 \oplus x^5$.

Specific values of coefficients $\delta_i(\tau)$ are uniquely defined by τ and the form of the generating polynomial $\varphi(x)$. They may be calculated by the algorithms proposed in [47]. Note that, regardless of the generating polynomial chosen, only one coefficient $\delta_{1+\tau}(\tau)$ in the set of expected coefficients $\{\delta_i(\tau)\}$ is 1 for $\tau < m$. Actually, as seen from Table 7.2, for $\tau = 1$ and 3, we obtain $\delta_1(1)\delta_2(1)\delta_3(1)\delta_4(1)\delta_5(1) = 01000$ and $\delta_1(3)\delta_2(3)\delta_3(3)\delta_4(3)\delta_5(3) = 00010$.

Any set of coefficients $\delta_i(\tau)$, $i = 1$ to m, determined by τ may be associated with the polynomial $\varphi_\tau(x) = \delta_1(\tau)x^0 \oplus \delta_2(\tau)x^1 \oplus \cdots \oplus \delta_{m-1}(\tau)x^{m-2} \oplus \delta_m(\tau)x^{m-1}$, which depends uniquely on specific values of $\delta_i(\tau)$. Polynomial $\varphi_\tau(x)$ can be obtained as the remainder of dividing x^τ by $\varphi^{-1}(x)$, which is conjugate to the generating polynomial $\varphi(x)$ [47]:

$$\varphi_\tau(x) = x^\tau[\text{mod } \varphi^{-1}(x)]. \tag{7.14}$$

Next we shall prove the theorem using the above definitions of polynomials $\varphi_\tau(x)$.

Theorem 7.4. The primitive polynomial $\varphi(x)$ used for constructing a test sequence generator allows one to produce all possible binary combinations in positions $Q_1, Q_2, \ldots, Q_n(Q_{j+1} > Q_j, j = 1$ to $n-1)$ of a shift register if and only if any subset of polynomials $\varphi_{\tau_0}(x) = x^{\tau_0}[\text{mod } \varphi^{-1}(x)]$, $\varphi_{\tau_1}(x) = x^{\tau_1}[\text{mod } \varphi^{-1}(x)], \varphi_{\tau_2}(x) = x^{\tau_2}[\text{mod } \varphi^{-1}(x)], \ldots$, where $\tau_{\gamma-1} = Q_\gamma - Q_1, \gamma = 1$ to n, is linearly independent over GF(2).

Proof. Suppose that the opposite statement is satisfied, i.e. all possible binary combinations will be produced in positions $Q_1, Q_2, \ldots, Q_n(Q_{j+1} > Q_j, j = 1$ to $n-1)$

of the shift register if polynomials $\varphi_{\tau_0}(x), \varphi_{\tau_1}(x), \varphi_{\tau_2}(x), \ldots, \varphi_{\tau_{n-1}}(x)$ or any set of them are linearly dependent over GF(2).

For the case of linear dependence of polynomials $\varphi_{\tau_0}(x), \varphi_{\tau_1}(x), \varphi_{\tau_2}(x), \ldots, \varphi_{\tau_{n-1}}(x)$, the equality

$$\varphi_{\tau_0}(x) \oplus \varphi_{\tau_1}(x) \oplus \varphi_{\tau_2}(x) \oplus \cdots \oplus \varphi_{\tau_{n-1}}(x) = 0 \tag{7.15}$$

is satisfied, which implies that, for coefficients $\delta_i(\tau_{\gamma-1}), i = 1$ to m, $\gamma = 1$ to n, defining the form of polynomial $\varphi_{\tau_{\gamma-1}}(x) = \delta_1(\tau_{\gamma-1})x^0 \oplus \delta_2(\tau_{\gamma-1})x^1 \oplus \cdots \oplus \delta_{m-1}(\tau_{\gamma-1})x^{m-1}$, the system of identities

$$
\begin{aligned}
&\delta_1(\tau_0) \oplus \delta_1(\tau_1) \oplus \delta_1(\tau_2) \oplus \cdots \oplus \delta_1(\tau_{n-1}) = 0 \\
&\delta_2(\tau_0) \oplus \delta_2(\tau_1) \oplus \delta_2(\tau_2) \oplus \cdots \oplus \delta_2(\tau_{n-1}) = 0 \\
&\quad\vdots \\
&\delta_m(\tau_0) \oplus \delta_m(\tau_1) \oplus \delta_m(\tau_2) \oplus \cdots \oplus \delta_m(\tau_{n-1}) = 0
\end{aligned}
\tag{7.16}
$$

holds true. Using (7.16), we may demonstrate that for the symbols of an M-sequence described by polynomial $\varphi(x)$, the equality

$$a(k + \tau_0) \oplus a(k + \tau_1) \oplus a(k + \tau_2) \oplus \cdots \oplus a(k + \tau_{n-1}) = 0 \tag{7.17}$$

where each symbol is represented by (7.12), holds true.

The equality (7.17) shows that the symbols of an M-sequence generated in the shift-register positions whose numbers are determined by $\{Q_1, Q_2, \ldots, Q_n\}$ are linearly dependent and hence all possible binary combinations cannot be produced for them. Their set is defined by expression (7.17), implying that the patterns $a(k + \tau_0)a(k + \tau_1)\,a(k + \tau_2)\cdots a(k + \tau_{n-1})$ consisting of an odd number of 1s cannot be produced.

We can similarly prove for any subset of polynomials $\varphi_{\tau_0}(x), \varphi_{\tau_1}(x), \varphi_{\tau_2}(x), \ldots, \varphi_{\tau_{n-1}}(x)$ that the statement contradicting the assumption of the theorem is inconsistent.

All possible binary combinations in shift-register positions $Q_1, Q_2, \ldots, Q_n$ will be provided only with no linear dependence between the symbols of an M-sequence generated in the shift register. Then, however, the system of identities (7.16) does not hold true for any subset of coefficients $\{\delta_i(\tau_{\gamma-1})\}$, $i = 1$ to m, $\gamma = 1$ to n, associated with a subset of symbols $a(k + \tau_{\gamma-1})$. Hence linear independence of polynomials $\{\varphi_{\tau_{\gamma-1}}(x)\}$ is ensured, which proves the theorem. ∎

Thus, prior to using the polynomial $\varphi(x)$ to create a test sequence generator for a VLSI circuit whose layout is described by the set of shift-register position numbers $\{Q_1, Q_2, \ldots, Q_n\}$, we need to investigate whether the conditions of Theorem 7.4 have been met. When the results are positive, all possible binary combinations, other than zero, are generated in positions $Q_1, Q_2, \ldots, Q_n$ of the VLSI shift register for the polynomial $\varphi(x)$ whose high degree is $m = n$. The latter is only true for $m > n$.

The analysis of the conditions of Theorem 7.4 is technically based on the use of one of the three relations (7.12), (7.15) and (7.16). In the first case, the linear relationship is tested for any subset of symbols in the M-sequence being analysed by representing them as a sum of symbols belonging to the set $\{a(k), a(k + 1), \ldots, a(k + m - 1)\}$. When the relation (7.15) is used, the algorithm for solving the above task is based on the procedure of binary polynomial division over GF(2) and, finally, for relation (7.16) an $m \times n$ matrix, is tested by evaluating its rank.

In spite of the close connection between the techniques that implement the above relations, their practical application may differ in effort and time required for proving the assumptions of Theorem 7.4. As an example of such a test, consider the digital circuit shown in Fig. 7.2, whose layout of connections between the combinational part and the shift register is described by the set $\{1, 4, 5\}$. Let the values of the M-sequence symbols generated by them be supported by polynomial $\varphi(x) = 1 \oplus x \oplus x^4$. Examining the set of shift-register position numbers $\{1, 4, 5\}$, we obtain $\tau_0 = 0$, $\tau_1 = 3$ and $\tau_2 = 4$, and their corresponding polynomials $\varphi_{\tau_0}(x) = x^0 [\bmod (1 \oplus x^3 \oplus x^4)] = x^0 = 1$, $\varphi_{\tau_1}(x) = x^3 [\bmod (1 \oplus x^3 \oplus x^4)] = x^3$, $\varphi_{\tau_2}(x) = x^4 [\bmod (1 \oplus x^3 \oplus x^4)] = 1 \oplus x^3$, where $\varphi^{-1}(x) = 1 \oplus x^3 \oplus x^4$. A simple test of expressions shows that for $\varphi_{\tau_0}(x)$, $\varphi_{\tau_1}(x)$ and $\varphi_{\tau_2}(x)$ the relation

$$\varphi_{\tau_0}(x) \oplus \varphi_{\tau_1}(x) \oplus \varphi_{\tau_2}(x) = 1 \oplus x^3 \oplus 1 \oplus x^3 = 0$$

holds true, thereby testifying to the impossibility of providing all possible binary combinations in the specified set $\{1, 4, 5\}$ of shift-register positions.

A similar result can be obtained by analysing the matrix of coefficients $\delta_1(\tau_0)\delta_2(\tau_0)\delta_3(\tau_0)\delta_4(\tau_0) = 1000$, $\delta_1(\tau_1)\delta_2(\tau_1)\delta_3(\tau_1)\delta_4(\tau_1) = 0010$, and $\delta_1(\tau_2)\delta_2(\tau_2)\delta_3(\tau_2)\delta_4(\tau_2) = 1010$:

$$A = \begin{vmatrix} 1 & 0 & 0 & 0 \\ 0 & 0 & 1 & 0 \\ 1 & 0 & 1 & 0 \end{vmatrix}.$$

The rank of the matrix is 2, which proves that the symbols of an M-sequence generated on the basis of coefficients $\delta_i(\tau_{\gamma-1})$, $i = 1$ to, 4, $\gamma = 1$ to 3, are linearly dependent.

The example discussed above is characterized by low cardinality of the set of register position numbers, and therefore its solution by any of the methods discussed will be trivial. The situation changes significantly when practical problems, for which the cardinality of the set of register position numbers normally exceeds 20, are to be solved. Special software which implements one of the earlier discussed methods is used for solving such problems. In a simple case, the software allows one to get a response to a question whether the polynomial under test is suitable for producing all possible binary combinations on a specified set of shift-register positions. If the response is negative, another primitive polynomial is tested for the purpose.

As has been reported in [114], the search for an appropriate polynomial is practically always reduced to a small number of primitive polynomials whose high degree m is n. The polynomial for which $m = n + 1$ may be used for solving the above problem only in exceptional cases.

More complex is the problem of finding a primitive polynomial $\varphi(x)$ that ensures exhaustive testing for the combinational VLSI part consisting of a set of subcircuits G_i, $i = 1$ to r, whose layout of connections to the shift register is described by the sets $\{Q_1^i, Q_2^i, \ldots, Q_{n_i}^i\}$. In this case the procedure of searching for an appropriate polynomial is iterative in nature. For the first subcircuit G_1 of the combinational VLSI part, we shall find a polynomial $\varphi(x)$ which provides for all possible combinations in positions $Q_1^1, Q_2^1, \ldots, Q_{n_1}^1$ of its shift register. Then, for the same polynomial, the remaining subcircuits G_j, $j = 2$ to r, of the VLSI circuit are exhaustively tested for input patterns. If the test has failed for a single set $\{Q_1^j, Q_2^j, \ldots, Q_{n_j}^j\}$, $j = 2$ to r, at least, the procedure of searching for a new polynomial is initiated for the first subcircuit G_1. Thus the process lasts until the polynomial $\varphi(x)$, which meets the assumptions of Theorem 7.4 for all sets $\{Q_1^i, Q_2^i, \ldots, Q_{n_i}^i\}$, $i = 1$ to r, is found.

In [114] is presented the primitive polynomial $\varphi(x) = 1 \oplus x^3 \oplus x^7 \oplus x^{10} \oplus x^{11} \oplus x^{12} \oplus x^{14} \oplus x^{18} \oplus x^{20}$, which exhaustively exercises the following sets of an 81-position shift register:

$$0, 1, 2, 3, 4, 5, 6, 7, 8, 9, 10, 11, 12, 13, 41, 42, 43, 44, 45, 46, 75$$
$$0, 1, 2, 3, 4, 5, 6, 13, 14, 15, 16, 17, 47, 48, 49, 50, 51, 76$$
$$0, 1, 2, 3, 4, 5, 6, 18, 19, 20, 21, 22, 52, 53, 54, 55, 56, 77$$
$$0, 1, 2, 3, 4, 5, 6, 23, 24, 25, 26, 27, 28, 57, 58, 59, 60, 61, 78$$
$$0, 1, 2, 3, 4, 5, 6, 29, 30, 31, 32, 33, 34, 63, 64, 65, 66, 67, 68, 79$$
$$0, 1, 2, 3, 4, 5, 6, 35, 36, 37, 38, 39, 40, 69, 70, 71, 72, 73, 74, 80.$$

The above methods of constructing test sequence generators for specified VLSI layout and their specific applications show that it is basically possible to implement the idea of VLSI self-testing by using pseudorandom sequences. As a major approach to putting the idea into practice, exhaustive testing has been proposed. An alternative approach may be random testing based on the use of pseudorandom sequences generated by primitive polynomials.

7.4. RANDOM TESTING OF VLSI CIRCUITS

Random testing of VLSI circuits that implement the LSSD technique in the test mode is based on generation of sequences of equally likely binary patterns. Such sequences may be generated by M-sequence generators, which require no significant overheads. Chapter 5 deals with estimating the efficiency of a specific generator structure. The procedure of obtaining information on whether the use of the random testing techniques discussed earlier is reasonably based on the complete description of the layout of the VLSI combinational part, which requires labour-consuming computations and does not allow one to analyse comparatively the characteristics of different M-sequence generators.

By analogy with exhaustive testing for VLSI circuits, we shall consider various possible approaches to measuring the efficiency of random testing for VLSI circuits only by the sets $\{Q_1^i, Q_2^i, \ldots, Q_{n_i}^i\}$, $i = 1$ to r, which describe the layout of connections between the combinational part and the memory elements in a VLSI circuit. Here the combinational part of the VLSI circuit will be represented as r subcircuits G_i, $i = 1$ to r, and the memory elements are connected as a shift register whose length w is determined by their number.

As a measure of random testing efficiency, let us introduce the probability $P_r(m, v)$ of producing all possible binary combinations in $v \leqslant m$ random positions of the shift register of length $w \geqslant m$.

By using an M-sequence described by the primitive polynomial $\varphi(x)$ we may prove the following theorem for probability $P_r(m, v)$.

Theorem 7.5. When an M-sequence described by $\varphi(x)$ for $2^m - 1 \geqslant w$, where $m = \deg \varphi(x)$, is generated, the probability $P_r(m, v)$ of producing all possible binary combinations in $v \leqslant n$ positions of the w-position shift register is defined by

$$P_r(m, v) = P(m, 3)P(m, 4)P(m, 5) \cdots P(m, v).$$

Here
$$P(m, t) = (2^m - 2^{t-1})/(2^m - t), t = 3 \text{ to } v, v \leqslant m.$$

Proof. Suppose that all possible binary combinations may be produced in the $(t-1)$th of v positions of the shift register. This is an indication of linear independence of M-sequence symbols associated with the $(t-1)$th register position. Estimate the probability that a randomly chosen tth position will form a linearly independent group of M-sequence symbols together with any subset of $(t-1)$th positions.

For the $(t-1)$th positions of the shift register, there are $2^{t-1} - 1$ non-empty subsets of positions and appropriate subsets of M-sequence symbols. Owing to the fact that the $(t-1)$th symbol of the M-sequence forms a linearly independent set, any pair of their subsets will also be linearly independent. Each of the 2^{t-1} non-empty subsets will be associated with only one distinct tth position such that an M-sequence symbol produced in it is linearly dependent on the original subset. An exception is the $(t-1)$ subsets consisting of a single shift-register position, due to the fact that the condition $w \leqslant 2^m - 1$ is satisfied. Hence the total number of occurrences when t symbols of the M-sequence are linearly dependent is

$$2^{t-1} - 1 - (t-1) = 2^{t-1} - t$$

and the number of positions where the tth symbol might be formed is $2^m - 1 - (t-1) = 2^m - t$. With the equally likely choice of the tth position in the shift register of length $w = 2^m - 1$, the probability $P(m, t)$ that the tth symbol of the M-sequence will be linearly independent of any subset from the $(t-1)$th preselected pattern may be obtained by

$$P(m, t) = \frac{2^m - t - 2^{t-1} + t}{2^m - t} = \frac{2^m - 2^{t-1}}{2^m - t} \qquad t \leqslant m. \qquad (7.18)$$

For $w < 2^m - 1$, the relation (7.17) may be used as the expected value of probability $P(m, t)$ and extended to the general case $w \leqslant 2^m - 1$.

The derivation of (7.17) is based on the assumption that the $(t-1)$th symbol of an M-sequence is as linearly independent as any subset of its symbols. Since the probability $P(m, t-1)$ of such an event is calculated by expression (7.17) and based on the linear independence of $(t-2)$ symbols and any of their subsets, the probability $P_r(m, v)$ of producing all possible binary combinations in v out of m positions of the shift register will be determined by the probability that all subsets of symbols in sets consisting of one, two, three, etc., and v symbols of the M-sequence are linearly independent at a time. Following from the relation $w \leqslant 2^m - 1$, we finally obtain

$$P_r(m, v) = P(m, 3)P(m, 4)P(m, 5) \cdots P(m, v) \qquad (7.19)$$

where $P(m, t)$, $t = 1$ to v, is calculated by (7.18) and $P(m, 1) = P(m, 2) = 1$, as was to be shown. ∎

The value of $P_r(m, v)$ is evidently a function of variables m and v, where $v \leqslant m$, and it depends uniquely on their specific ratio. We examine this dependence of $P_r(m, v)$ for $m = 10$ to 20 and the difference $m - v = 0$ to 5, whose estimates are summarized

Table 7.3.

m	\multicolumn{6}{c}{$m - v$}					
	0	1	2	3	4	5
10	0.31	0.60	0.80	0.91	0.96	0.98
11	0.30	0.59	0.79	0.90	0.95	0.98
12	0.29	0.59	0.78	0.89	0.95	0.98
13	0.29	0.58	0.78	0.89	0.94	0.97
14	0.29	0.58	0.77	0.88	0.94	0.97
$\vdots$	$\vdots$	$\vdots$	$\vdots$	$\vdots$	$\vdots$	$\vdots$
20	0.29	0.58	0.77	0.88	0.94	0.97

in Table 7.3. From the table it follows that, for practical values of $m > 14$, $P_r(m, m) = 0.29$. This implies that the use of primitive polynomial $\varphi(x)$ whose deg $\varphi(x) = m$ allows one to produce all possible binary combinations in $v = m$ specified positions of the VLSI shift register (see Fig. 7.1) with probability 0.29. To improve the probability of exhaustive coverage of the VLSI combinational part connected to v positions of the register, it is advisable to use the polynomials whose high degree $m > v$. Thus, for example, for $v = 10$ to 13, the dependence $P_r(m, v)$ shown in Fig. 7.7 demonstrates that even for $m = 14$ the probability of producing all possible binary combinations in the specified $v = 10$ positions of the VLSI shift register is 0.94 whereas it is 0.29 for $m = 10$.

Thus the basic outcome of Theorem 7.5 is not only the measurement of VLSI random testing efficiency but also the possibility of designing a test sequence generator based on selecting the high degree of generating polynomial $\varphi(x)$ on the basis of the confidence of VLSI exhaustive testing implementation. Furthermore, the probability $P_r(m, v)$ may be used for estimating the effort required for finding suitable primitive polynomial $\varphi(x)$ by the techniques of Sec. 7.3.

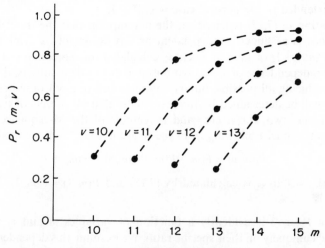

Fig. 7.7. Probability $P_r(m, v)$ versus m

The outcome of Theorem 7.5 may be interpreted as follows. A random primitive polynomial $\varphi(x)$ of degree m allows one to implement VLSI exhaustive testing with probability $P_r(m, v)$, where v is the number of inputs to the VLSI combinational part. This probability $P_r(m, v)$ implies that polynomial $\varphi(x)$ for exercising exhaustive coverage for the v-input combinational VLSI part will pass the test. In so doing, the search for an appropriate $\varphi(x)$ among the set of polynomials whose high degree is m will be characterized by rather low probability $P_r(m, m) = 0.29$. In this case, m is the number of inputs to a single subcircuit G_1 in the combinational VLSI part.

The solution of the problem for the general case characterized by exercising exhaustive coverage for n subcircuits of the VLSI circuit, whose maximal number of inputs does not exceed m, may be estimated by the probability of finding the desired polynomial $\varphi(x)$, i.e.

$$P_d = P_r^n(m, m) = 0.29^n. \tag{7.20}$$

Hence for $n > 10$ the probability of solving the polynomial search problem becomes practically zero. It should be noted that the expression (7.20) is nothing more than the estimate of the actual value of P_d. The obtained estimates of P_d can be significantly improved by means of several pseudorandom test sequence generators. Thus, for two test sequence generators, the probability P_d, which is similar to that defined by (7.20), will take the form

$$P_d = 1 - [1 - P_r^n(m, m)][1 - P_r^n(m, m)] = 2P_r^n(m, m) - P_r^{2n}(m, m).$$

From the estimates obtained, we may deduce that the use of several pseudorandom test sequence generators whose high degree m satisfies the condition $m > v$, where v is the maximal number of inputs to the subcircuits of VLSI circuit G, preferable. Then the upper bound of m will be determined by the test experiment time, which is proportional to $2^m - 1$.

As follows from the above discussion, the use of a set of pseudorandom sequence generators as well as the increase in the high degrees of their generating polynomials allow one to improve the effiiciency of random and exhaustive testing for VLSI chips.

7.5. RING TESTING

The implementation of so-called ring testing is the extension of pseudorandom test sequence use to VLSI self-testing. The idea behind ring testing is the use of M-sequence generators that form an autonomous circuit containing the combinational circuit under test. The structure has been examined in [119]. In the structure proposed, the shift register in conjunction with a modulo-2 adder inserted in a feedback loop forms an M-sequence generator whose input is applied with the symbols produced at the outputs of the combinational circuit (CC) under test.

The testing procedure for such a structure is implemented in the following three steps: initialization, check and analysis. In the initialization step, an initial code is stored in the shift register. Then shift control pulses are applied to the shift register in the checking step. Finally the code obtained at the shift register is compared with its reference value.

The most widely used structure for ring test implementation is the one where the

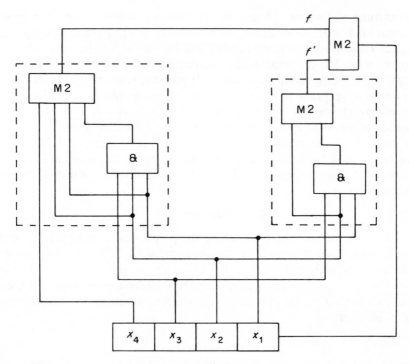

Fig. 7.8. A functional scheme to implement ring testing for a combinational circuit described by
$f = x_1 \oplus x_2 \oplus x_4 \oplus x_1 x_2 x_3$

circuit under test in conjunction with extra logic implements linear feedback for the
shift register [120]. As an example of the structure, consider ring testing for a
combinational circuit that implements a Boolean function $f = x_1 \oplus x_2 \oplus x_1 x_2 x_3 \oplus x_4$.
For the purpose, let us use the M-sequence generator described by polynomial
$\varphi(x) = 1 \oplus x \oplus x^4$. In order to apply linear feedback to the shift register, introduce
extra logic to implement the function $f' = x_2 \oplus x_1 x_2 x_3$. Then the modulo-2 sum
of f and f' will be defined by the expression $f \oplus f' = x_1 \oplus x_4$, which complies with
the selected primitive polynomial $\varphi(x)$. Figure 7.8 shows the functional scheme for
the example described.

The test coverage for the circuit of Fig. 7.8 as well as the appropriate characteristic
that implements ring testing are determined by the following factors:

(i) High degrees of primitive polynomial $\varphi(x)$, which describes the M-sequence
generator.

(ii) Layout of connections between the shift-register positions and the inputs to
the circuit under test.

(iii) Initial settings of the shift register.

(iv) The number of steps in the ring testing procedure.

The above considerations are closely related. Thus with the number of steps being
$2^m - 1$, where m is the high degree of generating polynomial $\varphi(x)$, the initial setting
of the shift register may only be zero. However, with fewer than $2^m - 1$ steps, the

initial setting becomes important. Therefore, in the general case, the problem of defining the major ring testing parameters is rather important.

To obtain the mentioned parameters for the structure that implements ring testing, we may use the analytical results presented in [119–121], which are based on the theory of signature analysis.

Let us discuss the basic concepts of the theory and its application to digital testing.

8

SIGNATURE ANALYSIS

8.1. THE CONCEPT OF SIGNATURE ANALYSIS

At present signature analysis is widely used in modern equipment for digital testing. The first signature analyser, HP5004A, was manufactured by Hewlett–Packard [65]. The purpose of the analyser is to detect errors in data streams caused by the faults of a digital device under test.

The conventional signature analyser scheme includes a shift register (1) and a modulo-2 adder (2) whose inputs are connected to the outputs of register positions in accordance with the generating polynomial $\varphi(x)$ (Fig. 8.1). The signature analyser control signals are START, CLOCK and STOP [1]. The START and STOP signals establish the time window within which data compression is performed by the analyser. The START signal resets all memory elements in the shift register (1) to zero and the shift register starts shifting bit by bit to the right by the CLOCK signal. On each CLOCK, the first position of the shift register is loaded with data expressed by (4.4)

$$a_1(k) = y(k) \oplus \sum_{i=1}^{m} \alpha_i a_i(k-1)$$

where $y(k) \in \{0, 1\}$ is the kth symbol of compressed sequence $\{y(k)\}, k = 1$ to l; $\alpha_i \in \{0, 1\}$ are the coefficients of generating polynomial $\varphi(x)$; $a_i(k-1) \in \{0, 1\}$ is the content of the ith memory element in the shift register in the $(k-1)$th cycle. The procedure of data shifting in the register is described by

$$a_j(k) = a_{j-1}(k-1) \qquad j = 2 \text{ to } m.$$

Thus the complete mathematical structure of signature analyser behaviour appears

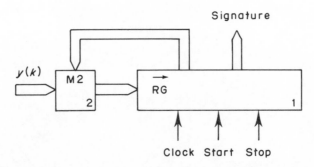

Fig. 8.1. Signature analyser block diagram: 1, shift register; 2, modulo-2 adder

as

$$a_i(0) = 0 \qquad i = 1 \text{ to } m$$

$$a_1(k) = y(k) \oplus \sum_{i=1}^{m} \alpha_i a_i(k-1) \tag{8.1}$$

$$a_j(k) = a_{j-1}(k-1) \qquad j = 2 \text{ to } m, \, k = 1 \text{ to } l$$

where l is normally taken to be equal to or less than $2^m - 1$, hence it determines the length of sequence to be compressed. When l cycles of signature analyser operation elapse, the binary code, which is a signature mapped into hexadecimal code, will be fixed on its memory elements.

Thus by generating a test sequence at the input lines to the digital device under test, we may find the expected signatures for each of its nodes, the set of which will be stored and used in the future for comparison with the actual signatures produced by devices under test. Any difference between the actual signature and the expected signature points to the fact that the circuit node behaves abnormally. In order to clarify the difference of signatures on the node, it is necessary to test the signatures successively from the specified node to the device inputs. This procedure is in many ways similar to that of fault detection in analogue devices, which consists of successive measurement and analysis of some analogue values. Such an approach determines the major advantage of signature analysis: its ease of use for fault detection and fault location in digital circuits, since it needs complex test hardware for carrying out the test experiment and minimal practical knowledge for its implementation. An example that illustrates the simplicity of signature analysis implementation is the functional scheme of the Hewlett–Packard analyser described by polynomial $\varphi(x) = 1 \oplus x^7 \oplus x^9 \oplus x^{12} \oplus x^{16}$. This analyser has been considered an unofficial standard to similar devices [122]. Its scheme comprises a five-input modulo-2 adder and a 16-position shift register as well as several gates for generation of START and STOP signals. The shift-register positions are grouped into four tetrads, whose contents define the value of hexadecimal signature by the coding table [122] (Table 8.1).

As a signature analyser example, let us consider the analyser described by the generating polynomial $\varphi(x) = 1 \oplus x^3 \oplus x^4$. Figure 8.2 shows the functional scheme of the analyser and Fig. 8.3 shows the timing chart of its operation, where the full arrows shows the transition from state $a_1(k-1)a_2(k-1)a_3(k-1)a_4(k-1)$ to new state $a_1(k)a_2(k)a_3(k)a_4(k)$ with $y(k) = 1$, and the broken arrows show the same for $y(k) = 0$. Consider the procedure of expected signature generation for the digital circuit of

Table 8.1.

0000	0	1000	8
0001	1	1001	9
0010	2	1010	A
0011	3	1011	C
0100	4	1100	F
0101	5	1101	H
0110	6	1110	P
0111	7	1111	U

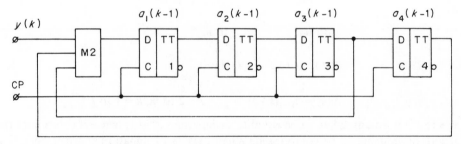

Fig. 8.2. A functional scheme of a signature analyser described by the polynomial $\varphi(x) = 1 \oplus x^3 \oplus x^4$

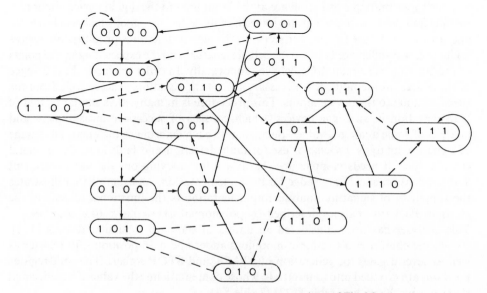

Fig. 8.3. Timing chart for the signature analyser of Fig. 8.2

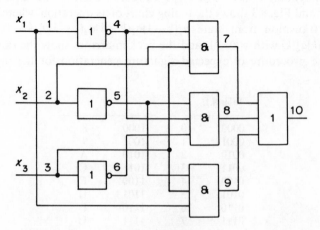

Fig. 8.4. A digital circuit

Table 8.2.

Test no.	Test sequence			Circuit node no.									
	x_1	x_2	x_3	1	2	3	4	5	6	7	8	9	10
1	0	0	0	0	0	0	1	1	1	0	1	0	1
2	0	0	1	0	0	1	1	1	0	0	0	0	0
3	0	1	0	0	1	0	1	0	1	0	0	0	0
4	0	1	1	0	1	1	1	0	0	1	0	0	1
5	1	0	0	1	0	0	0	1	1	0	0	0	0
6	1	0	1	1	0	1	0	1	0	0	0	1	1
7	1	1	0	1	1	0	0	0	1	0	0	0	0
8	1	1	1	1	1	1	0	0	0	0	0	0	0

Fig. 8.4 by applying the analyser of Fig. 8.2, which implements the relations (8.1). Suppose that exhaustive testing based on generation of all possible input patterns as a counter sequence is used to determine whether the circuit behaves normally. The expected responses to the applied input patterns are given in Table 8.2 for each node in the circuit.

The value of the expected signature for the sequence produced in the tenth node of the circuit is obtained by handling its symbols in accordance with the relation (8.1). The timing chart of the analyser's memory element states will appear as in Fig. 8.5. Any successive state here has been defined depending on the value of $y(k)$ (where $k = 1$ to 8) by the analyser behaviour chart (see Fig. 8.3). From the last transition for $y(8) = 0$, the final value of signature $S_{10} = 1011$ will be produced on the memory elements in the analyser. In accordance with the adopted coding option (see Table 8.1), we may write it down as $S_{10} = C$. The expected signature for other circuit nodes whose values have been summarized in Table 8.3 may be obtained similarly.

Since the expected signatures are indicated in the schematic diagram of the digital device in the same way as voltages and currents for analogue circuits, the procedure of fault detection in the circuit under test is significantly simplified.

Let us consider an example of fault testing and diagnosing for the circuit of Fig. 8.4, where a logic 0 level has been indicated on the fourth node due to a physical defect. The

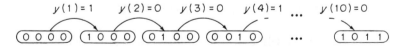

$y(1)=1 \qquad y(2)=0 \qquad y(3)=0 \qquad y(4)=1 \qquad \cdots \qquad y(10)=0$

Fig. 8.5. Timing chart of analyser's memory element states for the sequence produced in the tenth node of the circuit shown in Fig. 8.4

Table 8.3.

i	1	2	3	4	5	6	7	8	9	10
S_i	7	6	H	4	5	P	F	5	2	C

Table 8.4.

Test no.	Test sequence			Circuit node no.									
	x_1	x_2	x_3	1	2	3	4	5	6	7	8	9	10
1	0	0	0	0	0	0	0	1	1	0	0	0	0
2	0	0	1	0	0	1	0	1	0	0	0	0	0
3	0	1	0	0	1	0	0	0	1	0	0	0	0
4	0	1	1	0	1	1	0	0	0	0	0	0	0
5	1	0	0	1	0	0	0	1	1	0	0	0	0
6	1	0	1	1	0	1	0	1	0	0	0	1	1
7	1	1	0	1	1	0	0	0	1	0	0	0	0
8	1	1	1	1	1	1	0	0	0	0	0	0	0

truth table for the circuit with $f_4 = 0$ (Table 8.4) comprises exhaustive information on the circuit behaviour.

The first step of circuit analysis (see Fig. 8.4) is to determine the actual value of signature at its output. By compressing the sequence produced in the tenth node, we obtain $S_{10}^* = 2$, which is different from the expected value $S_{10} = C$. Thus, the difference between signatures S_{10}^* and S_{10} indicates that the circuit under test is faulty. Otherwise, with $S_{10}^* = S_{10}$, the hypothesis that the circuit under test is fault-free is accepted.

The second step of circuit analysis is to locate a fault. It is based on the principle of successive testing of signatures in the nodes that are functionally connected to the output node. This procedure is often referred to as back-tracing and for the example of interest it will involve successive testing of signatures in nodes 9, 8 and 7. As a result, we obtain $S_9^* = 2$ and $S_8^* = S_7^* = 0$; hence, only S_9^* equals the expected value. From further analysis of the values of S_4^*, S_5^* and S_6^*, we may state that $S_4^* = 0$ differs from its expected value $S_4 = 4$. Now, the equality $S_1^* = S_1 = 7$ as well as the match between the actual and expected signature for nodes 2, 3, 5 and 6 imply that the sought-for fault is caused by either the first NOT gate or its output node. This ends the procedure of analysing the digital circuit (see Fig. 8.4).

8.2. SIGNATURE ANALYSIS AS A BINARY POLYNOMIAL DIVISION ALGORITHM

Various mathematical models and algorithms have been used to describe the procedure of data compression based on primitive polynomials [123–125]. The model that is most commonly used implements the idea of data compression as polynomial division over GF(2) [123, 125]. The dividend may be considered as a data stream to be compressed that is represented as a polynomial $\kappa(x)$ of degree $l - 1$, where l is the number of bits in the sequence. Thus, for example, sequence 10011 may have the form of polynomial $\kappa(x) = x^4 \oplus x \oplus 1$. The divisor may be a primitive polynomial, and division results in a quotient $q(x)$ and remainder $S(x)$ that are related by the classical equation of the following form

$$\kappa(x) = q(x)\psi(x) \oplus S(x). \tag{8.2}$$

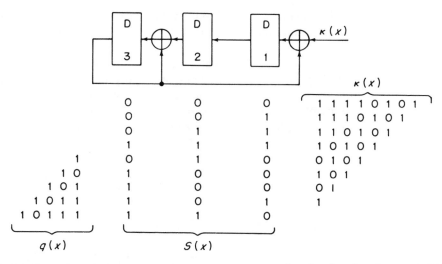

Fig. 8.6. Generation of signature $S(x)$ for data stream $\kappa(x) = x^7 \oplus x^6 \oplus x^5 \oplus x^4 \oplus x^2 \oplus 1$ and divisor $\psi(x) = x^3 \oplus x^2 \oplus 1$

The remainder $S(x)$, which is a polynomial of lower degree than $m = \deg \psi(x)$, is called a signature.

Figure 8.6 shows an example of a scheme generating a signature for data stream 11110101, described by polynomial $\kappa(x) = x^7 \oplus x^6 \oplus x^5 \oplus x^4 \oplus x^2 \oplus 1$ whose compression is performed by the analyser described by polynomial $\psi(x) = x^3 \oplus x^2 \oplus 1$. The initial state of memory elements in the signature analyser is assumed to be 0, although any specified state can be taken as an initial state in the general case. The data stream 11110101 to be compressed is successively applied to the analyser input, causing the memory elements to change their states as per the timing chart of Fig. 8.6. The remainder $S(x)$ of polynomial $\kappa(x) = x^7 \oplus x^6 \oplus x^5 \oplus x^4 \oplus x^2 \oplus 1$ divided by polynomial $\psi(x) = x^3 \oplus x^2 \oplus 1$ is set on the analyser's memory elements and assumes the value $S(x) = x^3 \oplus x$ as a polynomial or $S(x) = 110$ as a binary code of polynomial coefficients. The correspondence between $S(x)$ and the remainder of $\kappa(x)$ divided by $\psi(x)$ may be supported by the following example:

$$
\psi(x) = x^3 \oplus x^2 \oplus 1 \enclose{longdiv}{\begin{array}{l} x^4 \oplus x^2 \oplus x \oplus 1 = q(x) \\ x^7 \oplus x^6 \oplus x^5 \oplus x^4 \oplus x^2 \oplus 1 = \kappa(x) \\ \underline{x^7 \oplus x^6 \oplus x^4} \\ x^5 \oplus x^2 \\ \underline{x^5 \oplus x^4 \oplus x^2} \\ x^4 \\ \underline{x^4 \oplus x^3 \oplus x} \\ x^3 \oplus x \oplus 1 \\ \underline{x^3 \oplus x^2 \oplus 1} \\ x^2 \oplus x = S(x). \end{array}}
$$

Hence it follows that the remainder is equal to the value of signature being generated. A rigorous proof of the assumption follows uniquely from the theory of cyclic codes [85].

For practical implementation of the signature analyser described by polynomial $\psi(x) = x^3 \oplus x^2 \oplus 1$ (see Fig. 8.6), as well as for any other polynomial, there exists an alternative structure shown in Fig. 8.7, which is preferable from the hardware implementation standpoint [126]. However, as suggested in [127, 128], $C(x)$ resulting from data stream convolution by the signature analyser with modulo-2 adders does not coincide with the remainder, i.e. $C(x) \neq S(x)$. At the same time, there is a one-to-one correspondence between $C(x)$ and $S(x)$ that may be expressed in the general case as

$$
S(x) = \begin{vmatrix}
\alpha_m & 0 & 0 & \cdots & 0 & 0 \\
\alpha_{m-1} & \alpha_m & 0 & \cdots & 0 & 0 \\
\alpha_{m-2} & \alpha_{m-1} & \alpha_m & \cdots & 0 & 0 \\
\vdots & \vdots & \vdots & & \vdots & \vdots \\
\alpha_2 & \alpha_3 & \alpha_4 & \cdots & \alpha_m & 0 \\
\alpha_1 & \alpha_2 & \alpha_3 & \cdots & \alpha_{m-1} & \alpha_m
\end{vmatrix} C(x) \tag{8.3}
$$

where $C(x)$ results from convolution by the signature analyser described by polynomial $\varphi(x)$; $S(x)$ is the remainder of polynomial $\kappa(x)$ divided by polynomial $\psi(x)$, which is the inverse of $\varphi(x)$; and $\alpha_i \in \{0, 1\}$, $i = 1$ to m, are the coefficients of polynomial $\psi(x)$.

The validity of relation (8.3) stems from the fact that polynomial $\kappa(x) = x^l$ divided by a primitive polynomial results in binary coefficients $\delta_i(L - l)$, $i = 1$ to m, $L = 2^m - 1$, defining the connection layout for a multi-input modulo-2 adder on whose output an l-cycle-delayed M-sequence replica is formed [47]. On the other hand, to obtain the values of these coefficients we may simulate the operation of an M-sequence generator described by the appropriate inverse polynomial and

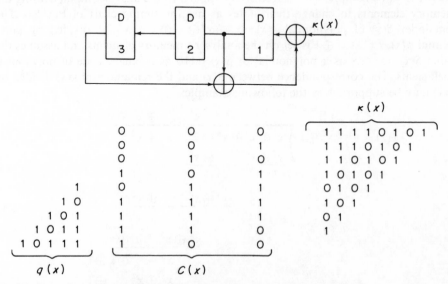

Fig. 8.7. Generation of signature $C(x)$ for an alternative analyser structure with $\kappa(x)$ $= x^7 \oplus x^6 \oplus x^5 \oplus x^4 \oplus x^2 \oplus 1$ and $\psi(x) = x^3 \oplus x \oplus 1$

multiply the simulation result by the array [47]

$$
\begin{vmatrix}
\alpha_m & 0 & 0 & \cdots & 0 & 0 \\
\alpha_{m-1} & \alpha_m & 0 & \cdots & 0 & 0 \\
\alpha_{m-2} & \alpha_{m-1} & \alpha_m & \cdots & 0 & 0 \\
\vdots & \vdots & \vdots & & \vdots & \vdots \\
\alpha_2 & \alpha_3 & \alpha_4 & \cdots & \alpha_m & 0 \\
\alpha_1 & \alpha_2 & \alpha_3 & \cdots & \alpha_{m-1} & \alpha_m
\end{vmatrix}.
$$

The linearity of modulo-2 addition proves that equation (8.3) will hold true not only for $\kappa(x) = x^l$, but also for arbitrary polynomials $\kappa(x)$.

For the specific case of Figs 8.6 and 8.7 signature $S(x)$ is obtained as a result of dividing $\kappa(x)$ by $\psi(x) = x^3 \oplus x^2 \oplus 1$, and $C(x)$ is obtained by compressing the input data stream on the analyser with modulo-2 adders described by $\varphi(x) = x^3 \oplus x \oplus 1$ where $\psi(x) = \varphi^{-1}(x)$. The relation (8.3) for $S(x) = 110$ and $C(x) = 100$ will have the form [129]

$$
S(x) =
\begin{vmatrix}
\alpha_3 & 0 & 0 \\
\alpha_2 & \alpha_3 & 0 \\
\alpha_1 & \alpha_2 & \alpha_3
\end{vmatrix}
C(x) =
\begin{vmatrix}
1 & 0 & 0 \\
1 & 1 & 0 \\
0 & 1 & 1
\end{vmatrix}
\times
\begin{vmatrix}
1 \\
0 \\
0
\end{vmatrix}
=
\begin{vmatrix}
1 \\
1 \\
0
\end{vmatrix}.
$$

Thus the examples of signature analysers and their mathematical models that have been presented here make it possible to scrutinize the procedure of signature generation and trace the relation between the theory of signature design and analysis and that of cyclic codes, M-sequence theory in particular. A major criterion that determines the wide application of signature analysis is its high efficiency, which can be estimated on the strength of some conditions stemming from the properties of M-sequences. Below we consider some techniques for estimating signature analysis effectiveness.

8.3. *SIGNATURE ANALYSIS EFFICIENCY*

Before examining signature analysis efficiency, it is interesting to note that reports on signature analysis application commonly do not disclose that the fault coverage in a digital circuit primarily depends on the quality of test stimuli. When the designated fault does not manifest itself in the output sequences of the circuit as a distortion of its symbols, it is undetectable by signature analysis, which is nothing more than an efficient data stream compression technique. Therefore if the data stream carries no information on the fault, neither will the compressed stream [31].

Thus by 'signature analysis efficiency' we shall basically mean its capability of error detection in the data stream being compressed. This signature analysis property may be estimated by various approaches and techniques. The most commonly used is the probabilistic approach, which consists of determination of probability P_n of failing to detect errors in the data sequence being analysed. Also, in the case being considered, only the probability dependent on the compression technique is estimated, whereas the other factors are ignored.

The value of P_n is calculated for a fairly general case that approximates to actual cases. It has been assumed that the reference data stream may equally likely take on different values and any error bit configuration may be an equally likely event. Next, by using the polynomial division algorithm as a body of mathematics for signature generation, we may demonstrate that for an l-bit dividend an $(l-m)$-bit quotient and an m-bit remainder (signature) are obtained. Then the correspondence between the actual l-bit sequence and the expected sequence is estimated by the equality of their m-bit signatures. An identical signature will be generated for 2^{l-m} distinct quotients. This implies that $2^{l-m}-1$ error l-bit sequences are associated with a single sequence, i.e. the reference one. Considering the equiprobability of error data streams, we may infer that $2^{l-m}-1$ error sequences initiating the reference signature are undetectable. Thus, the probability P_n of failing to detect errors in the data stream being analysed is calculated as

$$P_n = \frac{2^{l-m}-1}{2^l-1} \tag{8.4}$$

where 2^l-1 is the total number of error sequences.

The expression (8.4) for $l \gg m$ may be simplified to give

$$P_n \simeq 1/2^m \tag{8.5}$$

which may come as a convincing reason for the high efficiency of signature analysis. In fact, for the Hewlett–Packard signature analyser, $P_n = 0.000\,0152...$, which proves that the efficiency of the signature analyser is sufficiently high.

The value $1 - P_n = 0.999\,984...$ may be found in practically every report on signature analysis, but only some of the reports realistically estimate the relation, which prevents one from assessing the advantages of signature analysis. A simple example, which shows that the integral estimate (8.5) is unsuitable as a criterion of signature analysis efficiency compared with other compact testing methods, has been given in [126]. It consists of the fact that, with the use of a technique based on the analysis of only m of l bits by ignoring $l - m$ bits of the sequence under test and under the assumption adopted while deriving the relation (8.5), the probability P_n will also be calculated by the same formula. A similar example for the specific case of linear combinational circuits is presented in [130].

Along with the integral characteristic P_n, it has been shown in some cases that another signature analysis advantage is its ability to detect all single errors for any l and all double errors for $l \leqslant 2^m - 1$ [65, 122, 126]. The refined values of P_n, which have been given in [127], do not allow one to obtain a comparative signature analysis characteristic against other compact testing methods.

As a more accurate measure of signature analysis effectiveness, consider the dependence of the error escape probability on error multiplicity μ, i.e. find the value of P_n^μ, where $\mu = 1$ to $2^m - 1$. Here we shall use the fundamentals of the theory of M-sequence generation and take account of the fact that by linearity of modulo-2 addition polynomial $\kappa^*(x)$, which describes the error sequence, may be represented as

$$\kappa^*(x) = \kappa(x) \oplus e(x) \tag{8.6}$$

where $\kappa(x)$ is a polynomial that describes the reference sequence and $e(x)$ is the error polynomial. It is evident that the error escapes detection only if $e(x)$ is divisible by $\varphi(x)$, which describes the signature analyser structure [47].

If a single error occurs in the lth position of the binary data sequence, the error polynomial will have the form $e(x) = x^{l-1}$, where $1 \leqslant l \leqslant 2^m - 1$, $m = \deg \varphi(x)$ and $\varphi(x)$ is a primitive polynomial. Using the property of M-sequences to form an l-cycle-delayed M-sequence replica for m non-zero coefficients, we can demonstrate that every l is associated with a distinct remainder from dividing x^{l-1} by $\varphi(x)$. In other words, any single error described by polynomial $e(x) = x^{l-1}$ will be associated with a distinct non-zero signature $S(x)$. Thus the signature analyser based on the primitive polynomial $\varphi(x)$ will detect all single errors [126, 127].

When errors occur in the lth and rth positions of the binary data sequence, the error polynomial will have the form $e(x) = x^{l-1} \oplus x^{r-1}$. For the given polynomial, we can demonstrate that for any pair of integers $l > r$ ($1 \leqslant l \leqslant L$, $1 \leqslant r \leqslant L$, $L = 2^m - 1$) a non-zero signature is formed as a sum of two distinct non-zero signatures $e(x) = x^{l-1}$ and $e(x) = x^{r-1}$ over GF(2). In this case, linearity of summing over GF(2) is used.

Consequently, on the assumption that the length of the data sequence analysed does not exceed $2^m - 1$, the signature analyser based on $\varphi(x)$ detects all double errors, thereby supporting earlier results [65].

For L non-zero signatures calculated for error polynomials of the form $e(x) = x^{l-1} \oplus x^{r-1}$, there exists a set of polynomials $e(x) = x^{v-1}$ generating identical signatures. Therefore, with signatures for $e(x) = x^{l-1} \oplus x^{r-1}$ and $e(x) = x^{v-1}$ being identical, the signature obtained for $e(x) = x^{l-1} \oplus x^{r-1} \oplus x^{v-1}$ will be zero, causing no change in the resulting signature. In other words, triple error patterns defined by l, r and v are undetectable. In terms of M-sequence theory, the magnitudes of l, r and v are determined by the shift-and-add property (see Sec. 6.1). By using the property, one can demonstrate that for undetectable triple errors described by a polynomial $e(x) = x^{l-1} \oplus x^{r-1} \oplus x^{v-1}$, an identity $a(k+l-1) \oplus a(k+r-1) = a(k+v-1)$ is satisfied for M-sequence symbols $a(k+l-1)$, $a(k+r-1)$ and $a(k+v-1)$, when a signature analyser with generating polynomial $\varphi(x)$ is used. The generating polynomial for an M-sequence is then $\varphi(x)$. Hence the number of undetectable triple errors is determined by the number of identities consisting of three terms and satisfied for an M-sequence. For $L = 2^m - 1$ the number is [129]

$$V_n^3 = C_L^2 / 3 = (2^m - 1)(2^m - 2)/3! \tag{8.7}$$

where C_L^2 is the number of combinations of 2 out of $L = 2^m - 1$ and the presence of divisor 3 is explained by the fact that a single polynomial $x^{l-1} \oplus x^{r-1} \oplus x^{v-1}$ describing an undetectable error of three error bits is associated with three polynomials $x^{l-1} \oplus x^{r-1}$, $x^{l-1} \oplus x^{v-1}$ and $x^{r-1} \oplus x^{v-1}$ describing 2-bit errors.

The total number of possible triple error patterns is $V^3 = C_L^3$ and the probability of triple errors going undetected will be equal to the ratio of the undetected error count to their total

$$P_n^3 = V_n^3 / V^3 = 1/(2^m - 3).$$

The number of detected triple errors is determined by

$$V_d^3 = V^3 - V_n^3 = (2^m - 1)(2^m - 2)(2^m - 4)/3!. \tag{8.8}$$

For the detectable triple errors described by $e(x) = x^{l-1} \oplus x^{r-1} \oplus x^{v-1}$, a polynomial $e(x) = x^{w-1}$ can be found such that it can initiate an identical structure. Thus, $e(x) = x^{l-1} \oplus x^{r-1} \oplus x^{v-1} \oplus x^{w-1}$ describes an undetectable pattern of four invalid bits. The value of w is found by the shift-and-add property and is a function of variables l, r and v, i.e. $w = f(l, r, v)$. For $w \neq f(l, r, v)$, polynomial $e(x) = x^{l-1} \oplus x^{r-1} \oplus x^{v-1} \oplus x^{w-1}$ describes a detectable error of four invalid bits. The total number of undetectable 4-bit errors is defined subject to equation (8.8) by

$$V_n^4 = V_d^3/4 = (2^m - 1)(2^m - 2)(2^m - 4)/4!. \tag{8.9}$$

The presence of divisor 4 in equation (8.9), similar to divisor 3 in (8.7), is attributable to the fact that four polynomials $x^{l-1} \oplus x^{r-1} \oplus x^{w-1}$, $x^{l-1} \oplus x^{r-1} \oplus x^{v-1}$, $x^{l-1} \oplus x^{w-1} \oplus x^{v-1}$ and $x^{w-1} \oplus x^{r-1} \oplus x^{v-1}$ that describe detectable errors are associated with a single polynomial of the form $x^{l-1} \oplus x^{r-1} \oplus x^{v-1} \oplus x^{w-1}$ that describes an undetectable error.

Thus, the probability of a 4-bit error to go undetected is

$$P_n^4 = \frac{V_n^4}{V^4} = \frac{(2^m - 1)(2^m - 2)(2^m - 4)}{4! \, C_L^4} = \frac{1}{2^m - 3}.$$

By using a similar technique, we can demonstrate that the number of undetectable 5-bit errors is given by

$$V_n^5 = [V^4 - V_n^4 - V_n^3(2^m - 4)]/5.$$

The term $V_n^3(2^m - 4)$ suggests that any 5-bit error comprising an undetectable 3-bit error will always be detectable. Here $2^m - 4 = C_{L-3}^1$.

The number of undetectable 6-bit errors is

$$V_n^6 = [V^5 - V_n^5 - V_n^4(2^m - 5)]/6.$$

For the general case, we can demonstrate that the number of undetectable μ-bit errors with $\mu \in \{1, 2, 3, \ldots, 2^m - 1\}$ will be expressed by

$$V_n^1 = 0 \qquad V_n^2 = 0$$
$$V_n^\mu = [V^{\mu-1} - V_n^{\mu-1} - V_n^{\mu-2}(2^m + 1 - \mu)]/\mu. \tag{8.10}$$

Consider as an example the signature analyser of Fig. 8.6 for $L = 7$ to prove the validity of (8.10). From equation (8.10) we have evaluated V_n^μ for $\mu = 1$ to 7. The results are summarized in Table 8.5.

An experimental simulation of signature analyser operation for various binary error sequences allows us to find those for which $S(x) = 0$. Our simulation results are summarized in Table 8.6, where $e(x)$ is the polynomial describing an undetectable error and $q(x)$ is the quotient of $e(x)$ divided by $\varphi(x)$.

The analysis of results presented in Tables 8.5 and 8.6 proves the truth of (8.10), which suggests that for $m = 3$ there are seven polynomials describing 3-bit errors, seven polynomials associated with 4-bit errors and a single polynomial associated

Table 8.5.

μ	1	2	3	4	5	6	7
V_n^μ	0	0	7	7	0	0	1

Table 8.6.

$\varphi(x)$	$q(x)$	$e(x)$
1101	0001	0001101
1101	0010	0011010
1101	0100	0110100
1101	1000	1101000
1101	0111	0100011
1101	1110	1000110
1101	1101	1010001
1101	0011	0010111
1101	0110	0101110
1101	1100	1011100
1101	0101	0111001
1101	1010	1110010
1101	1101	1100101
1101	1111	1001011
1101	1011	1111111

with a 7-bit error that divide evenly by polynomial $\varphi(x)$ for deg $\varphi(x) = 3$. Hence seven 3-bit, seven 4-bit and one 7-bit errors will go undetected.

Owing to the possibility of determining the dependence of the number of undetectable errors on their multiplicity (8.10), we may derive the escape probability distribution for errors of any multiplicity. In this case the error escape probability will be expressed as [47].

$$P_n^1 = 0 \qquad P_n^2 = 0$$

$$P_n^\mu = \frac{V^{\mu-1} - V_n^{\mu-1} - V_n^{\mu-2}(2^m + 1 - \mu)}{\mu V^\mu} \qquad \mu = 3 \text{ to } L \quad (8.11)$$

where $V^\mu = C_L^\mu$. By rearranging equation (8.11) we finally obtain

$$P_n^1 = P_n^2 = 0$$

$$P_n^\mu = \frac{1}{2^m - \mu}[1 - P_n^{\mu-1} - (\mu - 1)P_n^{\mu-2}] \quad \mu = 3 \text{ to } 2^m - 1. \quad (8.12)$$

On the basis of (8.12), for $\mu = 3, 4, 5, 6, 7$ we obtain

$$P_n^3 = \frac{1}{2^m - 3} \qquad P_n^4 = \frac{1}{2^m - 3} \qquad P_n^5 = \frac{2^m - 8}{(2^m - 5)(2^m - 3)}$$

$$P_n^6 = \frac{2^m - 8}{(2^m - 5)(2^m - 3)} \qquad P_n^7 = \frac{(2^m - 3)(2^m - 5) - 7(2^m - 8)}{(2^m - 3)(2^m - 5)(2^m - 7)}.$$

From the analysis of the resulting P_n^μ and equation (8.12), it becomes evident that $P_n^\mu \simeq 1/2^m$, $\mu = 3$ to $2^m - 2$, for sufficiently large m. Thus, μ-multiple errors with $\mu \in \{3, 4, 5, ..., 2^m - 2\}$ that might occur in a data sequence of length $2^m - 2$ are detected by the signature analyser with probability $P_d^\mu = 1 - P_n^\mu$, which is practically 1 (100%) with $m > 7$.

An important deduction from expression (8.12) is the fact that it is invariant as to the compressed sequence and the form of primitive polynomial $\varphi(x)$. The only parameter that effects the values of P_n^μ is the high degree m of $\varphi(x)$.

8.4. ANALYSIS OF COMPARATIVE SIGNATURE ANALYSIS EFFICIENCY ESTIMATION METHODS

The signature data compression technique was compared analytically against other compact testing techniques for the first time in [65]. The present analysis is based on estimating the probability P_n of failing to detect errors in the data stream for the two compact testing methods, i.e. signature analysis and transition counting. The estimation of P_n is carried out for both techniques under sufficiently general assumptions. The first assumption is that any reference data sequence is an equally probable event. The second assumption is that any error bit configuration may occur with equal probability.

Under these assumptions, as has been shown in Sec. 8.3, the value of P_n may be obtained by (8.4), and it is $1/2^m$, where m is the signature length, for long enough sequences. For the transition count technique, the value of P_n is also calculated as the ratio of all undetectable errors in the data stream of length l to all possible errors. However, the number of undetectable errors cannot be determined in a straightforward manner. This is due to the non-linearity of the transition count operation and hence the dependence of the number of undetectable errors on the form of the reference sequence. In fact, the number of undetectable errors in the sequence with r transitions may be defined by the expression [65]

$$C_l^r - 1 = \frac{l!}{r!(l-r)!} - 1$$

and the probability of an event such that a sequence with r transitions will be generated may be defined by

$$C_l^r/2^l.$$

Thus, the average number of errors undetectable by transition counting is defined under the above assumptions by

$$\sum_{r=0}^{l} \frac{(C_l^r - 1)C_l^r}{2^l}.$$

The relation between the latter expression and the total number of possible sequences characterizes the probability P_n:

$$P_n = \frac{\sum_{r=0}^{l} (C_l^r - 1)C_l^r/2^l}{2^l - 1}.$$

Considering that $C_l^r = l!/r!(l-r)!$ and assuming that l takes on high enough values, we obtain [65]

$$P_n \simeq 1/(\pi l)^{1/2}. \tag{8.13}$$

For $l = 2^m$, which corresponds to the length of m-bit checksums, we finally obtain

$$P_n \simeq 1/(\pi 2^m)^{1/2}.$$

Hence the probability of failing to detect an error by transition counting is sufficiently higher than that for signature analysis. Apart from (8.13), a major reason for the use of signature analysis is the probability P_n^1 of failing to detect single errors, which is $1/2$ for transition counting and 0 for signature analysis [65]. Thus by comparing the relations (8.4) and (8.13) as well as the estimates for P_n^1, we may conclude that signature analysis is more advantageous over other compact testing methods [65].

However, a slight variation in the rather general initial assumptions may produce the opposite result. Thus when it is assumed that generation of sequences with r transitions, where $r \in \{0, 1, 2, ..., l\}$, is an equallly likely event, the probability P_n will be calculated as [131]

$$P_n = \frac{1}{l+1} \sum_{r=0}^{l} \frac{C_l^r - 1}{2^l - 1} \tag{8.14}$$

where $C_l^r - 1$ is the number of error sequences consisting of r transitions.

At the same time, for a particular digital circuit the number r of transitions in its reference response is known in advance, and the same probability P_n will appear as

$$P_n = \frac{C_l^r - 1}{2^l - 1} \tag{8.15}$$

for transition counting.

Therefore, apart from the estimate of P_n as per (8.13), we may use other estimates, for example (8.14) and (8.15), and compare them against the appropriate value for signature analysis. In so doing, consider that the value of P_n for signature analysis is invariant relative to the sequence compressed. Hence it follows, in particular, that the expression (8.4) also holds true for the assumptions adopted when the relations (8.14) and (8.15) were derived. Owing to this the given compact testing methods may be compared under other similar conditions. Thus for the case of equiprobability of generating a sequence with r transitions we may demonstrate that the ratio of P_n for the two methods will be 1 with $l = 2^m - 1$. This suggests that transition counting and signature analysis techniques are equivalent as to their efficiency measure represented as the fault escape probability P_n. This example is a good illustration of the fact that forced assumptions may lead to quite different conclusions on the effectiveness of the same techniques.

The comparison between signature analysis and transition count testing for the actual case with known transition count r in the sequence under test makes it possible to estimate the probability ratio P_n as

$$\frac{2^{l-m} - 1}{C_l^r - 1}. \tag{8.16}$$

It is evident that for different r values the ratio (8.16) varies over a wide range, with signature analysis being more efficient for values less than 1; otherwise, transition counting is used. Now we consider some examples demonstrating the efficiency of

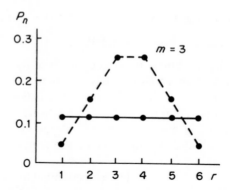

Fig. 8.8. Fault escape probability P_n for transition counting (----) and signature analysis (———) with $m = 3$

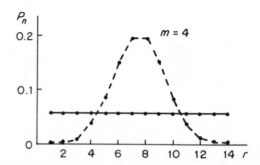

Fig. 8.9. Fault escape probability P_n for transition counting (----) and signature analysis (———) with $m = 4$

signature analysis and transition counting as a function of r. Let $l = 2^m - 2$ and the appropriate probability values P_n for both techniques be calculated by expressions (8.4) and (8.15). For $m = 3$ and $m = 4$ we obtain the plots of Figs 8.8 and 8.9. The full line shows the probability P_n plot for signature analysis, and the broken 'curve' shows the plot for transition counting.

From the plots obtained we may deduce that with $m = 3$ the signature analysis efficiency estimate is greater than that of transition counting only for $r = 2, 3, 4$ and 5 transitions in the data stream being analysed. For $r = 1$ and 6, transition counting is more efficient. At the same time, with $m = 4$, the number of r's for which the latter outperforms signature analysis increases substantially. In fact, signature analysis is only efficient for $r = 5, 6, 7, 8, 9$ and 10, which is the lesser portion of all possible r's.

The plot of (8.16) against $r/(2^m - 2)$ makes it evident that the range of r for which signature analysis efficiency exceeds that of transition counting is narrower. The range becomes narrower with the increase in the tested sequence length, which is $2^m - 2$ in this case (Fig. 8.10).

When using the relation (8.16) for estimating signature analysis efficiency, the following things should be considered. First, the expression (8.16) may only be used for the case when the number of transitions r is known in the reference sequence.

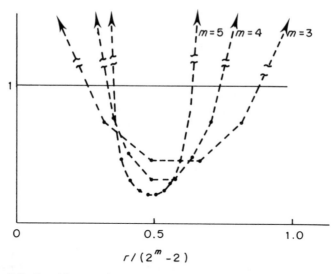

Fig. 8.10. Transition counting (----) versus signature analysis (———) efficiency

Secondly, such comparative analysis is only true for the hypothesis that the error bit sequences caused by a circuit failure are equally likely.

The relation (8.16), when used for deriving more general conclusions on signature analysis efficiency, may produce invalid results. Thus, we may deduce that transition counting is superior to signature analysis since for sufficiently long sequences l the range of efficient r is narrow for signature analysis. Moreover, we may easily show that for $m \to \infty$ the range of variable $r/(2^m - 2)$ for which the relation (8.16) is less than 1 tends to the same value of 0.5. This may be explained by the fact that the number of r-transition sequences changes with r, which is not accounted for by (8.16). In fact, from the quick analysis of plots in Fig. 8.9 we may conclude that, with $m = 4$, transition counting is more suitable since it is more efficient for $r = 1, 2, 3, 4, 11, 12, 13$ and 14 whereas signature analysis is efficient only for six values out of 14. However, the total number of sequences $C_{15}^1 + C_{15}^2 + C_{15}^3 + C_{15}^4 + C_{15}^{11} + C_{15}^{12} + C_{15}^{13} + C_{15}^{14}$ for which the transition count technique is preferable turns out to be less than the number of sequences C_{15}^5 that consist for $r = 5$ transitions only. This fact has been taken into account in deriving the relation (8.13).

Thus by estimating the signature analysis efficiency we came to the conclusion that any assumption of equiprobability of any error sequence, the same as any other general assumption, prevents us from obtaining even an approximate signature analysis estimation compared to other compact testing methods. The principal consideration in comparing the methods is the presence of more detailed characteristics of compressed sequences associated with their possible errors in particular.

The first approach to estimating the advantages of signature analysis on the basis of a more complete description of possible errors in the sequence being compressed has been proposed in [126]. Here two classes of errors are considered: random errors and errors with invalid symbols appearing with a distance that equals the degree 2. In our opinion, the latter occur more often in practice. The mentioned publication reports on estimating the signature analysis efficiency for these error types and shows that the errors that occur in a sequence being compressed are dependent on each other. However, an attempt has been made here to prove the universal nature of signature analysis, its usability for any digital circuit applied with random test sequences in the test mode.

A more universal way of comparing compact testing techniques, which takes account of tested sequence properties and allows one to define the areas of signature analysis application, has been proposed in [132]. Let us consider its basic concepts.

8.5. AN APPROACH TO COMPARING COMPACT TESTING METHODS

The principle of the approach is that the efficiency of any compact testing method may be estimated by using the error escape probability P^μ, which depends on the error multiplicity $\mu \in \{1, 2, 3, ..., l\}$, where l is the length of sequence to be tested. The indicated error escape probability for multiplicity μ may be obtained by [132]

$$P^\mu = P_v^\mu P_n^\mu \tag{8.17}$$

where P_v^μ is the probability of an error of multiplicity μ; P_n^μ is the probability of failing to detect an error of multiplicity μ, defined as the ratio of undetectable μ-multiple errors to the number of all possible errors of μ error symbols in the sequence of length l.

The value P_v^μ in the expression (8.17) is determined by the form of circuit under test, the set of possible errors in it, as well as by the type of test sequences. Then the probability distribution P_v^μ may be absolutely random and vary heavily, depending on the error occurring, circuit configuration and test sequence. At the same time, the probability P_n^μ of failing to detect a μ-multiple error is characterized by the compact testing method only. Therefore, we can obtain the measure of circuit test efficiency as distribution P^μ (8.17) for different methods depending on their probability distribution P_n^μ. Based on an analysis of this distribution we may choose a more suitable method out of several compact testing methods. To simplify the algorithm for decision making, it is advisable that a more compact characteristic, e.g. the total error escape probability P_n calculated as the arithmetic sum of component values P^μ

$$P_n = \sum_{\mu=1}^{l} P^\mu = \sum_{\mu=1}^{l} P_v^\mu P_n^\mu \tag{8.18}$$

be used.

In this case, the value P_n characterizes one or other compact testing method for specific probability distribution P_v^μ of a fault depending on its multiplicity rather than any hypothetical assumption.

By way of example, consider an approach to comparing compact testing methods by using signature analysis, ones counting and a trivial method based on choosing any m and l tested bits as a signature. First of all, let us estimate the probability distribution P_n^μ for each of the listed methods.

For the case of signature analysis, the probability distribution P_n^μ is determined by the relation (8.12) and hence takes the form of Fig. 8.11 for $m = 4$, 5 and 6, respectively.

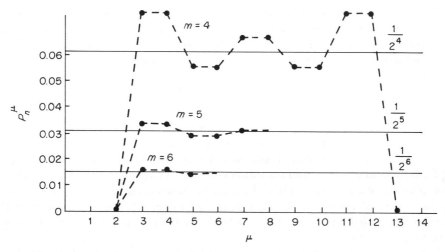

Fig. 8.11. Probability distribution P_n^μ of failing to detect a μ-multiple error by signature analysis

From the figure and the simulation results summarized in Table 8.7, one can see that with m as low as 7 the probability distribution P_n^μ may be thought of as practically uniform for signature analysis, and the values of P_n^μ for $\mu \in \{3, 4, 5, ..., 2^m - 2\}$ as equal to $1/2^m$.

Thus, we may conclude that the errors of multiplicity $\mu \in \{3, 4, 5, ..., 2^m - 2\}$ that have occurred in the data stream of length $2^m - 1$ for $m > 7$ will be detected by signature analysis with equal probability of $1 - 1/2^m$.

It should be noted then that single and double errors are detected with probability 1 and an error consisting of $2^m - 1$ error bits remains undetectable.

In the case of the 1s counting method, the signature value is determined by the 1s count r in the tested sequence, which may assume values from 0 to l, where l is taken to be $2^m - 1$ for comparability of results. It is evident then that for $\mu = 2k + 1$, $k = 0$ to $2^{m-1} - 1$, i.e. for odd values of μ, all errors can be detected by ones counting since any change in the odd number of symbols always involves a change in the signature [133]. Since the value of the latter remains unchanged, the error will be undetectable only when any change in one of its component signatures is compensated for by another. For the sequence of r 1s, these is a possibility for $\mu \leq 2r$-multiple errors, where $\mu = 2k$ ($k = 0$ to r, $r = 0$ to $2^{m-1} - 1$), to go undetected. Then a μ-multiple error that has occurred in the sequence under test will be undetectable only when $\mu/2$ 1s

Table 8.7.

m	$1/2^m$	$\max\|1/2^m - P_n^\mu\|$	$\dfrac{\max\|1/2^m - P_n^\mu\|}{1/2^m}$
3	0.1181	0.0819	0.69
4	0.0625	0.0144	0.23
5	0.0312	0.0032	0.10
6	0.0156	0.0007	0.04
7	0.0078	0.0002	0.03

and $\mu/2$ 0s in the sequence under test are inversed due to the fault. The total number of such situations for the specified μ and r will be $C_r^{\mu/2} C_{2^m-1-r}^{\mu/2}$, and the number of all possible μ-multiple errors is $C_{2^m-1}^{\mu}$. Then the probability P_n^{μ} of failing to detect μ-multiple errors in the tested sequence of r 1s will be defined by the expression [133]

$$P_n^{\mu} = \frac{C_r^{\mu/2} C_{2^m-1-r}^{\mu/2}}{C_{2^m-1}^{\mu}} \qquad \mu = 2k, \; k = 1 \text{ to } w \qquad (8.19)$$

where $w = r$ if $r \in \{1, 2, 3, ..., 2^{m-1} - 1\}$ or $w = 2^m - 1 - r$ if $r \in \{2^{m-1}, 2^{m-1} + 1, 2^{m-1} + 2, ..., 2^n - 2\}$, and the value of P_n^{μ} for $\mu = 0$ and $\mu = 2^m - 1$ is zero.

The analysis of the expression obtained has shown that the value of probability P_n^{μ} for both ones counting and transition counting is a function that depends on the multiplicity μ of a possible error and the reference sequence form, i.e. the 1s count r in the sequence. Therefore, the dependence of values P_n^{μ}, which characterize the distribution probability of failing to detect μ-multiple errors, is selected from the dependences for various r.

As an example, let us obtain the family of probability distributions for $r = 1, 2$ and 3 by equation (8.19). For $r = 1$, only P_n^2 is non-zero

$$P_n^2 = \frac{C_1^1 C_{2^m-2}^1}{C_{2^m-1}^2}.$$

For $r = 2$, non-zero values are assumed by probabilities

$$P_n^2 = \frac{C_2^1 C_{2^m-3}^1}{C_{2^m-1}^2} \qquad P_n^4 = \frac{C_2^2 C_{2^m-3}^2}{C_{2^m-1}^4}.$$

for $r = 3$

$$P_n^2 = \frac{C_3^1 C_{2^m-4}^1}{C_{2^m-1}^2} \qquad P_n^4 = \frac{C_3^2 C_{2^m-4}^2}{C_{2^m-1}^4} \qquad P_n^6 = \frac{C_3^3 C_{2^m-4}^3}{C_{2^m-1}^6}.$$

The dependence based of probability distribution P_n^{μ} for ones counting with the

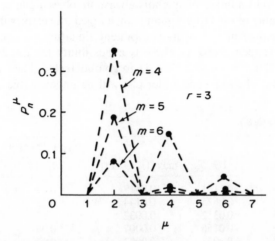

Fig. 8.12. Probability distribution P_n^{μ} of failing to detect a μ-multiple error by 1s counting

reference sequence consisting of $r = 3$ 1s and containing $l = 2^4 - 1$, $2^5 - 1$ and $2^6 - 1$ symbols are shown in Fig. 8.12. The analysis of these dependences reveals the non-uniformity of probability distribution P_n^μ, which increases with the value of m defining the length l of sequence to be compressed.

For a method based on selection and analysis of any m of l tested bits as a signature, the number of undetectable errors of multiplicity μ is determined as C_{l-m}^μ and the total number of possible errors by C_l^μ. Then the value of probability P_n^μ for $l = 2^m - 1$ takes on the form

$$\frac{C_{2^m-1-m}^\mu}{C_{2^m-1}^\mu} = \frac{(2^m - 1 - m)!(2^m - 1 - \mu)!}{(2^m - 1 - m - \mu)!(2^m - 1)!} \tag{8.20}$$

where $\mu = 1$ to $2^m - 1 - m$. For $\mu \in \{2^m - m, 2^m - m + 1, \ldots, 2^m - 1\}$, the probability P_n^μ is zero.

The dependence of probability distribution P_n^μ for $m = 4$, 5 and 6 shown in Fig. 8.13 reveals their non-uniformity for $m \to \infty$. Thus, the maximum difference P_n^μ is

$$\max \Delta P = P_n^1 - P_m^{2^m-m} = \frac{2^m - 1 - m}{2^m - 1}.$$

The latter expression is practically 1 for sufficiently large m.

Thus the probability distributions P_n^μ (8.12), (8.19) and (8.20) obtained allow one to compare the three compact testing methods on the basis of expression (8.18), and to substantiate the suitability of one of them for testing a specific digital circuit. As an example of such analysis, consider the digital circuit of Fig. 8.4. Let us determine which of the compact testing methods is most efficient in detecting faults $\equiv 0$ and $\equiv 1$ on all the circuit nodes. Assume then that the probability for two or more stuck-at faults to occur at a time is 0 and any of the specified faults is equally likely. Next, let a counter sequence of $l = 2^4 - 1 = 15$ repeatable 3-bit patterns be a test sequence.

First of all, let us find the probability distribution P_v^μ of an error of multiplicity $\mu = 1$ to 15. Thus we obtain output responses of the circuit (see Fig. 8.4) for all its faulty modifications. The behaviour of any fault modification of the circuit will be

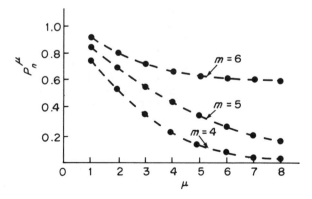

Fig. 8.13. Probability distribution P_n^μ of failing to detect a μ-multiple error by selection and analysis of a signature of any m out of l tested bits

described by an analytical expression derived from a Boolean function that has been implemented by the circuit of Fig. 8.4. This function has the form $F = \bar{x}_1 x_2 x_3 + \bar{x}_1 \bar{x}_2 \bar{x}_3 + x_1 \bar{x}_2 x_3$ and hence, for fault $f_1 \equiv 0$ where $f_1 = x_1$, is the function implemented by the first node of the circuit whose mathematical model is $F^* = x_2 x_3 + \bar{x}_2 \bar{x}_3$. The switching functions for all possible circuit faults are calculated similarly and summarized in the truth table for all faulty modifications of the circuit. The complete truth table is presented as Table 8.8.

Comparison studies of the reference circuit response with the faulty circuit responses make it possible to find the sets of errors caused by the faults. Thus, the analysis of a circuit with fault $x_1 \equiv 0$ reveals that the fault causes a five-fold error. In this way, the table for V_v^μ possible errors in the output circuit response is constructed versus their multiplicity μ (Table 8.9).

Based on data summarized in the table, we may plot the probability distribution $P_v^\mu = V_v^\mu / V$ as a function of μ (Fig. 8.14). There the total number of errors ($V = 20$) is taken into account. It is evident from the plot of P_v^μ that the distribution obtained differs markedly from the classical ones, thereby supporting once again that any assumptions for P_v^μ are questionable.

By using the values P_v^μ, $\mu = 1$ to 15, we may estimate the efficiency of each compact testing method that has been discussed earlier by (8.18). Then for signature analysis

Table 8.8.

			Reference output response	Output response of digital circuit																		
				With fault $\equiv 0$ at the node										With fault $\equiv 1$ at node									
x_1	x_2	x_3		1	2	3	4	5	6	7	8	9	10	1	2	3	4	5	6	7	8	9	10
0	0	0	1	1	1	1	0	0	0	1	0	1	0	0	0	0	1	1	1	1	1	1	1
0	0	1	0	0	0	0	0	0	0	0	0	0	0	1	1	0	0	0	1	1	1	1	1
0	1	0	0	0	1	0	0	0	0	0	0	0	0	0	0	1	0	1	0	1	1	1	1
0	1	1	1	1	0	0	0	1	1	0	1	1	0	0	1	1	1	1	1	1	1	1	1
1	0	0	0	1	0	1	0	0	0	0	0	0	0	0	0	1	1	0	0	1	1	1	1
1	0	1	1	0	1	0	1	0	1	1	1	0	0	1	0	1	1	1	1	1	1	1	1
1	1	0	0	0	0	0	0	0	0	0	0	0	0	0	0	0	0	0	0	1	1	1	1
1	1	1	0	1	1	0	0	0	0	0	0	0	0	0	0	0	1	1	1	1	1	1	1
0	0	0	1	1	1	1	0	0	0	1	0	1	0	0	0	0	1	1	1	1	1	1	1
0	0	1	0	0	0	0	0	0	0	0	0	0	0	1	1	0	0	0	1	1	1	1	1
0	1	0	0	0	1	0	0	0	0	0	0	0	0	0	0	1	0	1	0	1	1	1	1
0	1	1	1	1	0	0	0	1	1	0	1	1	0	0	1	1	1	1	1	1	1	1	1
1	0	0	0	1	0	1	0	0	0	0	0	0	0	0	0	1	1	0	0	1	1	1	1
1	0	1	1	0	1	0	1	0	1	1	1	0	0	1	0	1	1	1	1	1	1	1	1
1	1	1	0	0	0	0	0	0	0	0	0	0	0	0	0	0	0	0	0	1	1	1	1

Table 8.9.

μ	1	2	3	4	5	6	7	8	9	10	11	12	13	14	15
V_n^μ	0	4	3	2	2	5	0	0	4	0	0	0	0	0	0

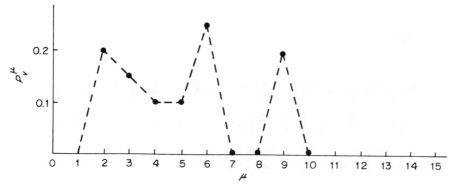

Fig. 8.14. Probability distribution P_v^μ as a function of fault multiplicity μ for the circuit of Fig. 8.4

we obtain

$$P_n = \sum_{\mu=1}^{15} P_v^\mu P_n^\mu = 0.051. \tag{8.21}$$

For 1s counting, considering that P_v^μ is zero for odd μ, we obtain

$$P_n = \sum_{\mu=1}^{15} P_v^\mu P_n^\mu = P_v^2 P_n^2 + P_v^4 P_n^3 + P_v^6 P_n^6 = 0.101.$$

Based on (8.20) we shall have $P_n = 0.202$ for the trivial method.

Hence we may conclude that, for the example under study, signature analysis is more efficient, with the probability of detecting the faults of Table 8.8 being maximal, i.e. $P_d = 1 - P_n = 0.949$. More accurate measures of compact testing method efficiency may be obtained for the example in question with due account of the number of errors dependent on their multiplicity μ and on the error configuration. However, such analysis as well as accurate estimation of P_v^μ is impracticable for real digital circuits. Therefore, the probability distribution P_v^μ is normally determined by simulating the limited set of faults or by *a priori* data on the test sequence. Thus, for the case of a test sequence whose patterns are designed for detecting disjoint fault sets, we may assume that the value of P_v^1 will be maximum and P_v^μ will be practically zero for $\mu = 2, 3, \ldots$. For the case when random and pseudorandom sequences are used as test stimuli, distribution P_v^μ is most likely to be uniform.

Let us consider some of the salient features of signature analysis as applied to testing digital circuits.

9

SPECIAL FEATURES OF SIGNATURE ANALYSIS APPLICATION

9.1. SIGNATURE ANALYSIS APPLICATION AREAS

Signature analysis is one of the most widely used compact testing methods at present. The principle of signature analysis is that long output responses of digital circuits are represented as short keywords called signatures. On the basis of their analysis the state of the circuit under test is estimated.

Signature analysis is used for implementing random, pseudorandom and exhaustive testing, since these methods of performing a test experiment are characterized by very long test sequences (10^6 patterns).

It is evident that the use of signature analysis in digital testing systems based on deterministic test design methods is not very efficient since the length of output circuit responses is normally not over 10^6. Furthermore, the test efficiency will be accordingly reduced due to failing to detect some error bit configurations in the output circuit responses by signature analysis. Then the probability P_d of detecting a fault may be defined by

$$P_d \simeq P_p(1 - P_n) \tag{9.1}$$

where P_p is the percentage of fault detection probability by violation of its output response which determines the test sequence coverage; P_n is the probability of failing to detect the tested sequence violations by signature analysis.

Thus signature analysis is more promising as an output response compression method to implement the random, pseudorandom or exhaustive testing for which the probability characteristic is the more accurate in describing the faulty state of the circuit. Therefore, by using the probability distribution P_v^μ of μ-multiple errors to occur in the output responses of the circuit, we may estimate the efficiency of each of the discussed methods on the basis of how the compact testing method is selected (see Sec. 8.5). Since it is very difficult and in most cases impracticable to obtain a sufficiently accurate description of probability distribution P_v^μ, we shall try to estimate the signature analysis efficiency under the following rather loose assumptions.

Various faults that might occur in the digital circuit affect its output response. Moreover, the probability distributions both for circuit faults and for the errors in the output sequences take a random form depending on their multiplicity. Thus for sequences of length $2^m - 2$, where m is the keyword width, probabilities P_v^μ, $\mu = 1$ to

$2^m - 2$, which assume random values, are defined as

$$P_v^\mu = V_v^\mu / V \tag{9.2}$$

where V_v^μ is the number of possible error sequence containing error bits and V is the total number of error sequences

$$V = \sum_{\mu=1}^{2^m-2} V_v^\mu. \tag{9.3}$$

By using the compact testing efficiency measurement technique, we shall determine the maximal and minimal values of probability P_n (8.18) for three of the methods discussed earlier.

For signature analysis, we consider that $P_n^\mu = 1/2^m, \mu = 3$ to $2^m - 2$, for rather long sequences, and $P_n^1 = P_n^2 = 0$. To obtain the maximal value of P_n, suppose that P_v^μ takes a random form for $\mu = 3$ to $2^m - 2$ and $P_v^1 = P_v^2 = 0$. Then

$$\max P_n = \sum_{\mu=1}^{2^m-2} P_v^\mu P_n^\mu = \sum_{\mu=3}^{2^m-2} P_v^\mu \frac{1}{2^m} = \frac{1}{2^m} \sum_{\mu=3}^{2^m-2} P_v^\mu.$$

Subject to equations (9.2) and (9.3), we finally obtain

$$\max P_n = 1/2^m. \tag{9.4}$$

The minimal estimate of P_n will be found for the case when the set V consists of only two subsets V_v^1 and V_v^2. Then we obtain

$$\min P_n = \sum_{\mu=1}^{2^m-2} P_v^\mu P_n^\mu = 0.$$

For the 1s counting method, the worst case estimated by $\max P_n$ is the case when $P_v^2 = 1$ and $P_n^1 = P_n^\mu = 0$ for $\mu = 3$ to $2^m - 2$. Based on (8.19) we find

$$\max P_n = P_v^2 P_n^2 = P_n^2 = \frac{C_r^1 C_{2^m-2-r}^1}{C_{2^m-2}^2} \tag{9.5}$$

where $r = 1$ to $2^{m-1} - 1$. The maximum value of relation (9.5) may be obtained with $r = 2^{m-1} - 1$. Then the given expression is transformed to

$$\max P_n = \frac{C_{2^{m-1}-1}^1 C_{2^m-2-2^{m-1}+1}^1}{C_{2^m-2}^2} = \frac{(2^{m-1}-1)(2^m-2^{m-1}-1)2}{(2^m-2)(2^m-3)} = \frac{2^{m-1}-1}{2^m-3}. \tag{9.6}$$

The minimal value of P_n for the 1s counting method may be obtained at $P_v^\mu = 0$ for $\mu = 2k, k = 1$ to $2^{m-1} - 1$, and hence assumes zero value for random probability distribution $P_v^\mu, \mu = 2k + 1, k = 0$ to $2^{m-1} - 2$.

For the method based on selection and analysis of any m of $2^m - 2$ tested bits as a signature, the worst case, which determines the maximal value of P_n, is when $P_v^1 = 1$ and $P_v^\mu = 0$ for $\mu = 2$ to $2^m - m - 2$. Then, by (8.20) we obtain

$$\max P_n = P_v^1 P_n^1 = P_n^1 = \frac{(2^m-2-m)!(2^m-3)!}{(2^m-3-m)!(2^m-2)} = \frac{2^m-m-2}{2^m-2} = 1 - \frac{m}{2^m-2}. \tag{9.7}$$

The minimal estimate of P_n for the latter method, in the same way as for the preceding two, is also zero when $P_v^\mu = 0$ at $\mu = 1$ to $2^m - 2 - m$. The given estimates of P_n as well as their limiting values are summarized in Table 9.1.

Analysis of the results obtained shows that signature analysis turns out to be more efficient in practically every case when data stream length is large, since its maximum P_n is characterized by the value $1/2^m$ and is practically zero even for $m > 7$. In so doing, the value of P_n does not exceed its maximum estimate for any form of probability distribution P_v^μ.

Let us calculate the value of P_n by (8.18) for the widely used assumption on equiprobability of any error bit configuration, for which P_v^μ is defined by

$$P_v^\mu = C_l^\mu/(2^l - 1)$$

and the value of P_n^μ by (8.11). Then we obtain

$$P_n = \sum_{\mu=1}^{l} P_v^\mu P_n^\mu = \sum_{\mu=1}^{l} \frac{C_l^\mu}{2^l - 1} \frac{V_n^\mu}{C_l^\mu} = \frac{1}{2^l - 1} \sum_{\mu=1}^{l} V_n^\mu$$

where the sum of the number of errors V_n^μ that are undetectable depending on their multiplicity is the total number of all undetectable errors and equals $2^{l-m} - 1$ for the signature analyser of length m. then we may finally obtain $P_n \simeq 1/2^m$, which corresponds to the maximally possible value of P_n.

It should be noted that, in some cases, signature analysis compares unfavourable with other compact testing methods regardless of its high efficiency. Thus, for example, for reference sequences appearing as $000\ldots0$ and $111\ldots1$, the ones counting method provides for detection of all possible errors and hence the zero value of P_n. On the contrary, the use of signature analysis will lead to missing error bit configurations whose number is determined by (8.10) depending on the error multiplicity. We may also give examples where the ones counting method, the same as any other compact testing method, will be more advantageous than signature analysis. A major advantage of signature analysis compared with other compact testing methods is the uniformity of probability distribution P_n^μ, which ensures a small fault undetection probability P_n for any form of P_v^μ. This explains the widespread use of signature analysis in practical implementations of test and diagnostic procedures for digital circuits.

Table 9.1.

Technique	min P_n	max P_n	lim max P_n $m \to \infty$
Signature analysis	0	$\dfrac{1}{2^m}$	0
Ones counting	0	$\dfrac{2^{m-1} - 1}{2^m - 3}$	$\dfrac{1}{2}$
Trivial	0	$1 - \dfrac{m}{2^m - 2}$	1

9.2. SIGNATURE ANALYSIS VALIDITY IMPROVEMENT TECHNIQUES

As has been mentioned earlier, the most exhaustive characteristic of signature analysis is the probability distribution P_n^μ of failing to detect an error, depending on its multiplicity $\mu = 1$ to l, where l is the length of sequence that is tested by this method, and for large l, P_n^μ is practically $1/2^m$ at any μ, where $m = \deg \varphi(x)$ and $\varphi(x)$ is a primitive generating polynomial used for signature analysis implementation. Moreover, the values of P_n^μ do not depend on the form of primitive polynomial $\varphi(x)$ but are defined by the value of its high degree m. Therefore one of the techniques of improving signature analysis efficiency is that based on the increase of m. However, even at $m = 16$ the value of the expression $1/2^{16} = 0.000\,0152\ldots$ is fairly close to zero and signature analysis efficiency does not manifest any tendency to increase with m. On the other hand, the increase of m causes an undesirable extension of signatures. Therefore signature analysers described by primitive polynomial $\varphi(x)$ whose high degree does not exceed 32 as a rule are more suitable in practice. The most often used value of m is 16.

The same as the above approach to improving signature analysis efficiency, other methods tend to vary the values of P_n^μ quantitatively. The results produced, however, may be quite different. One of the most frequently used requirements imposed on signature analysis is to satisfy the condition $P_n^\mu = 0$ for the specified range of μ whose execution is normally confined to the first members in the number series $\mu = 0, 1, 2, \ldots, 2^m - 1$. Thus for $P_n^1 = P_n^2 = 0$, it is sufficient to use the primitive polynomial $\varphi(x)$ for which the relations (8.11) determining the form of probability distribution P_n^μ have been satisfied. However, even for $P_n^1 = P_n^2 = P_n^3 = 0$, a number of solutions may exist.

A simple solution based on the use of primitive polynomials is the method that implements the signature analysis property. The principle of the method is that the number of undetectable errors depending on their multiplicity is determined by the high degree of polynomial $\varphi(x)$ whereas their configuration is closely related to the form of $\varphi(x)$.

Thus we may use some signature analysers implemented by primitive polynomials with identical high degree whose sets of undetectable errors with specified multiplicity are disjoint. An example of such a solution is the use of the corollary to Theorem 7.1, which states that the sets of undetectable errors of multiplicity $\mu = 3$ by polynomials $\varphi(x)$ and $\varphi^{-1}(x)$ are disjoint if and only if $L \bmod 3 \neq 0$, where $L = 2^m - 1$, $m = \deg \varphi(x)$ and $\varphi^{-1}(x)$ is the conjugate of $\varphi(x)$.

The creation of a signature analyser that detects all single, double and triple errors may be demonstrated by the use of polynomials $\varphi_1(x) = 1 \oplus x \oplus x^3$ and $\varphi_2(x) = 1 \oplus x^2 \oplus x^3$, for which the following relations hold true: $\varphi_1(x) = \varphi_2^{-1}(x)$ and $(2^3 - 1) \bmod 3 \neq 0$. When applied separately, each of the above polynomials ensures that the signature analyser based on the polynomial detects all single and all double errors. By (8.7) none of the analysers under consideration detects triple errors

$$V_n^3 = \frac{(2^3 - 1)(2^3 - 2)}{3!} = 7.$$

Table 9.2.

$\varphi_1(x)$	$\varphi_2(x)$
0001101	0001011
0011010	0010110
0110100	0101100
1101000	1011000
0100011	0110001
1000110	1100010
1010001	1000101

As seen from Table 9.2, the sets of undetectable errors from both analysers are disjoint.

Thus when analysing sequences of length $l \leqslant 2^m - 1$ by two analysers that have been based on polynomials $\varphi(x)$ and $\varphi^{-1}(x)$ meeting the assumptions of Theorem 7.1, we may detect all errors of multiplicity $\mu = 1, 2, 3$. Here the result is obtained by doubling the signature length, which is $2m$. We may also demonstrate that the mechanical applicatiion of primitive polynomial $\varphi(x)$ whose $\deg(x) = 2m$ will not produce the same result. Thus an error of the form 0100011 cannot be detected by the analyser described by polynomial $\varphi(x) = 1 \oplus x^5 \oplus x^6$ whereas it can be detected by two analysers described by polynomials $\varphi_1(x)$ and $\varphi_2(x)$ (Table 9.2). By employing such an approach we may also significantly decrease the value of P_n^μ for errors of multiplicity greater than 3. This is explained by the fact that the sets of undetectable errors of the specified multiplicity are never identical as a rule.

By using m signature analysers described by one and the same polynomial $\varphi(x)$, we may attain fat better results. In this case, each of the analysers handles only the specified set of symbols from the sequence under test. For this reason, an error that has occurred in the initial sequence is represented as errors of different configurations for any of m analysers.

Thus the actual error sequence has the form of a set of simulated sequences, each of which may contain a different pattern of error bits. If the error bit configuration is detectable by at least one analyser described by polynomial $\varphi(x)$, then the error occurring will be detectable by analysing all m signatures. A simple example of the initial sequence partitioning is the use of code characters that identify the numbers of elements in the sequence under test.

An example of such partitioning for a sequence of length $l = 2^4 - 1 = 15$ is given in Table 9.3.

The sequence under test is represented as four sequences of 0s and symbols $y(k) \in \{0, 1\}, k = 1$ to 15. Each of the sequences formed is associated with its own signature analyser described by polynomial $\varphi(x)$ whose high degree is 4. Each signature analyser detects errors depending on their multiplicity in accordance with (8.10), which ensures detectability of all single and double errors as noted above.

Suppose an error of five invalid symbols $y(4), y(6), y(11), y(12)$ and $y(13)$ that has occurred in the initial sequence is represented by violation of three, four, two and two characters respectively in the four newly formed sequences. Hence this error will always be detectable since it is represented as double errors for the third and fourth sequences, which are also detectable.

Table 9.3.

Sequence element no.		Tested sequence $\{y(k)\}, k = 1$ to 15	Simulated sequences			
Decimal code	Binary code		1	2	3	4
1	0001	$y(1)$	0	0	0	$y(1)$
2	0010	$y(2)$	0	0	$y(2)$	0
3	0011	$y(3)$	0	0	$y(3)$	$y(3)$
4	0100	$y(4)$	0	$y(4)$	0	0
5	0101	$y(5)$	0	$y(5)$	0	$y(5)$
6	0110	$y(6)$	0	$y(6)$	$y(6)$	0
7	0111	$y(7)$	0	$y(7)$	$y(7)$	$y(7)$
8	1000	$y(8)$	$y(8)$	0	0	0
9	1001	$y(9)$	$y(9)$	0	0	$y(9)$
10	1010	$y(10)$	$y(10)$	0	$y(10)$	0
11	1011	$y(11)$	$y(11)$	0	$y(11)$	$y(11)$
12	1100	$y(12)$	$y(12)$	$y(12)$	0	0
13	1101	$y(13)$	$y(13)$	$y(13)$	0	$y(13)$
14	1110	$y(14)$	$y(14)$	$y(14)$	$y(14)$	0
15	1111	$y(15)$	$y(15)$	$y(15)$	$y(15)$	$y(15)$

In the general case, any error occurring in the initial sequence may be represented as a set of errors of lower multiplicity whose cardinality M_1 is estimated by

$$\mu \geqslant M_1 \geqslant \text{int} \log_2 \mu$$

where μ is the multiplicity of error that has occurred. Then for a triple error $\mu = 3$ we obtain $M_1 \geqslant 2$. Hence any triple error will be detectable since it is represented by two errors of multiplicity 2 or 1, to say the least. We may similarly demonstrate detectability of all errors of multiplicity 4, 5, etc. It should be noted that the maximal multiplicity μ of detectable errors depends on m in the general case.

A major disadvantage of the above approach is the increase of signature length, which equals m^2 in the present case. At the same time, the proposed method does not require a complex design procedure and allows one to attain high detectability by using codes with high code distance as the numbers in the initial sequence.

Apart from the discussed methods of constructing signature analysers, which are capable of detecting errors with the specified multiplicity, there are some commonly used methods based on the theory of cyclic codes. Thus, for example, we may detect all single, double and triple errors by using the extended Hamming code or errors with multiplicity $\mu \leqslant 6$ by the Golay code. Moreover, based on the theory of cyclic codes, one can construct a signature analyser that is capable of detecting an error with the specified multiplicity μ.

9.3. SIGNATURE ANALYSIS APPLICATION FOR TROUBLESHOOTING

The use of signature analysis for diagnosing digital circuits makes it possible to detect a fault as well as to locate the fault and sometimes to establish the type of fault that has occurred. Consider some features of applying signature analysis for troubleshooting a digital circuit. As an example, analyse the sequences generated at the nodes of logic element F used for its implementation.

In the general case, logic element F can represent an arbitrary logic function and in the course of testing the circuit, which employs the element, test data sequences are produced at its nodes. The length l of the sequence is normally defined as $l \leqslant 2^m - 1$ where m is the high degree of generating polynomial $\varphi(x)$ used for creating the signature analyser.

Examine the dependence between the value of signature produced by the analyser that implements polynomial $\varphi(x)$ and the type of fault that has occurred. Let us first examine an example of a stuck-at-0 fault at one of the specified element nodes. In this case, the sequence of symbols $y(k) = 0$, $k = 1$ to l, will be generated at the faulty node of the circuit that comprises gate F in the process of digital circuit testing. According to (8.2), the value of signature $S(x)$ for a zero sequence will also be associated with he zero code, i.e. $S(0) = 00 \ldots 0$. Therefore when the actual signature $S^*(x) = S(0)$ differs from the expected signature $S(x)$, we may assume that a stuck-at-0 fault has occurred on this node.

When a stuck-at-1 fault occurs, a sequence of symbols $y(k) = 1$, $k = 1$ to l, formed on the faulty node of gate F may be described by polynomial $\kappa^1(x) = 1 \oplus x \oplus x^2 \oplus \cdots \oplus x^{l-3} \oplus x^{l-2} \oplus x^{l-1}$, which consists of the modulo-2 sum of polynomials of the form $\kappa_k(x) = x^{k-1}$, $k = 1$ to l.

Using the representation of unit sequence $\{y(k)\}$ as polynomial $\kappa^1(x)$, let us prove the following theorem.

Theorem 9.1. The value of signature $S(1)$ for a unit sequence of symbols $y(k) = 1$, $k = 1$ to $2^m - 1$, where $m = \deg \varphi(x)$ and $\varphi(x)$ is a primitive polynomial used for constructing the signature analyser, equals the signature of zero sequence of symbols, i.e. $S(1) = S(0)$.

Proof. Let us represent the original unit sequence $\{y(k)\}$ as a polynomial

$$\kappa^1(x) = 1 \oplus x \oplus x^2 \oplus \cdots \oplus x^{l-3} \oplus x^{l-2} \oplus x^{l-1} \tag{9.8}$$

which consists of the modulo-2 sum of polynomials $\kappa_k(x) = x^{k-1}$, $k = 1$ to l and $l = 2^m - 1$. Either of the above polynomials $\kappa_k(x)$ may be written as a relation (8.2):

$$\kappa_k(x) = q_k(x)\varphi^{-1}(x) \oplus S_k(x) \qquad k = 1 \text{ to } 2^m - 1. \tag{9.9}$$

Substituting the values of $\kappa_k(x)$ into expression (9.8) we obtain

$$\kappa^1(x) = \varphi^{-1}(x) \sum_{k=1}^{2^m-1} q_k(x) \oplus \sum_{k=1}^{2^m-1} S_k(x)$$

where the second term is the signature $S(1)$ for the unit sequence of symbols $y(k) = 1$, $k = 1$ to $2^m - 1$,

$$S(1) = \sum_{k=1}^{2^m-1} S_k(x). \tag{9.10}$$

The expression (9.10) is a bit-by-bit sum modulo-2 of all possible m-bit codes except for a zero code since, for $l = 2^m - 1$, the values of signatures $S_k(x), k = 1$ to l, form a complete set of non-zero m-bit signatures (see Sec. 8.3).

Thus the final value of $S(1)$ equals a zero code, i.e. $S(1) = S(0)$ as we wished to prove. ∎

By using the results of the above theorem, we can demonstrate that a stuck-at fault occurs at the node under test when the condition $S^*(x) = S(0)$ is satisfied and the expected value of $S(x)$ differs from that of $S(0)$.

To classify the fault that has occurred it is necessary to analyze the signature for any $l < 2^m - 1$. In this case, the inequality $S(1) \neq S(0)$ is most probably satisfied, thereby allowing one to state which of the two possible faults $\equiv 0$ or $\equiv 1$ has occurred eventually. The value $S(1)$ for $l < 2^m - 1$ can be obtained directly from (8.2) or as a result of an application experiment, whereas, for $l = 2^m - 1$ and any polynomial $\varphi(x)$ of degree m, $S(1) = 00\ldots00$.

When an inverse fault occurs the sequence $\{y(k)\}$ generated at any node of a logic element will be the inverse of reference sequence $\{y(k)\}, k = 1$ to l. Then the value of the actual signature $S^*(x)$ for the specified node will be defined as $S^*(x) = S(x) \oplus S(1)$, which follows from the theorem below.

Theorem 9.2. The value of signature $S(x)$ for the tested sequence $\{y(k)\}, k = 1$ to l, equals the modulo-2 sum of signatures $S^*(x)$ and $S(1)$ for the inverse sequence $\{\overline{y(k)}\}$ and the sequence consisting of all 1s, respectively.

Proof. The tested sequence $\{y(k)\}, k = 1$ to l, may be represented as a bit-by-bit modulo-2 sum of the inverse sequence $\{\overline{y(k)}\}$ and the sequence consisting of all 1s. Then for polynomials $\kappa(x)$ and $\overline{\kappa(x)}$ which describe the sequences $\{y(k)\}$ and $\{\overline{y(k)}\}$, respectively, and the polynomial $\kappa^1(x)$ of (9.8), the relation

$$\kappa(x) = \overline{\kappa(x)} \oplus \kappa^1(x) \qquad (9.11)$$

is satisfied.

Using the relations (8.2) and (9.11), we may demonstrate that the actual value of signature $S(x)$ will be determined by

$$S(x) = S^*(x) \oplus S(1) \qquad (9.12)$$

as we wished to prove. ∎

The value of signature $S^*(x)$ for the node of a circuit with an inverse fault equals the bit-by-bit modulo-2 sum of the expected signature $S(x)$ and the signature $S(1)$ of the unit sequence. For $l = 2^m - 1$, the specified fault will be signature-undetectable since in this case $S(1) = 00\ldots0$ and hence $S^*(x) = S(x)$ according to (9.12).

The values of signatures for the three faults discussed above, which are sometimes [5] used to build a complete model of a faulty digital circuit, are summarized in Table 9.4, where the value of expected signature $S(x)$ is represented by the binary code $a_1(l)a_2(l)\ldots a_m(l)$, and the value of $S(1)$ by the code $a_1^1(l)a_2^1(l)\ldots a_m^1(l)$. From the above results it is evident that the most suitable length l of the tested sequence is $2^m - 2$ to ensure detectability of the inverse fault and different values of signatures for faults $\equiv 0$ and $\equiv 1$.

Table 9.4.

Length of tested sequence	Fault type		
	$\equiv 0$	$\equiv 1$	Inverse
$l < 2^m - 2$	$00\ldots0$	$a_1^1(l)a_2^1(l)\ldots a_m^1(l)$	$a_1(l)a_2(l)\ldots a_m(l)$ $\oplus\quad\oplus\quad\oplus$ $a_1^1(l)a_2^1(l)\cdots a_m^1(l)$
$l = 2^m - 2$	$00\ldots0$	$00\ldots1$	$a_1(l)a_2(l)\cdots a_m(\overline{l})$
$l = 2^m - 1$	$00\ldots0$	$00\ldots0$	$a_1(l)a_2(l)\ldots a_m(l)$

The presence of a bridging fault between the nodes of gate F is indicated by the equality of actual signatures for the specified nodes.

The faults occurring at the nodes of memory elements employed in sequential digital circuits are defined in a similar way. However, the problem of attaining a suitable fault localization level is rather complex in the case of sophisticated circuit-level designs with feedback loops. As an example, consider a sequential circuit consisting of a memory element implemented by the D flip-flop and two-input EOR, OR, AND and NOT gates (Fig. 9.1). For testing the behaviour of the circuit, an M-sequence generator implemented by polynomial $\varphi(x) = 1 \oplus x^3 \oplus x^4$ is used. Here we shall use $2^4 - 2$ sequential values of four symbols of the M-sequence, the first three of which are applied as test sequences to the three data inputs 2, 4 and 5. First of all, the D flip-flop is set to 0 by a control signal C_1 and the test pattern is changed by a control signal C_2. Considering the initial conditions, the expected sequences (Table 9.5) are produced at all data lines of the circuit under test.

The values of expected signatures $S_i(x), i = 2, 4, 5, 6, 7, 8, 9, 10$, produced by the signature analyser (see Fig. 8.2) are shown in the logic diagram of the device (Fig. 9.1). The coding table (see Table 8.1) has been used to generate these signatures.

Consider the localization procedure for a 'mix-up' type fault, which consists of using the two-input AND gate instead of the two-input OR gate in the above circuit. In this case the sequences of Table 9.6 will be produced at the data lines of the circuit.

By sequentially analysing the signatures, the set of their actual values $S_i^*(x)$ (Table 9.7) can be obtained on the data lines of the circuit.

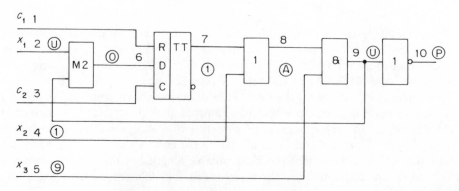

Fig. 9.1. A sequential circuit with feedback loop

Table 9.5.

Test pattern no.	Test sequence			Sequences at circuit nodes							
	x_1	x_2	x_3	2	4	5	6	7	8	9	10
1	1	0	0	1	0	0	1	0	0	0	1
2	1	0	1	1	0	1	0	1	1	1	0
3	1	1	1	1	1	1	0	0	1	1	0
4	0	0	0	0	0	0	0	0	0	0	1
5	0	0	1	0	0	1	0	0	0	0	1
6	0	1	0	0	1	0	0	0	1	0	1
7	1	1	1	1	1	1	0	0	1	1	0
8	0	0	1	0	0	1	0	0	0	0	1
9	0	1	1	0	1	1	1	0	1	1	0
10	1	0	1	1	0	1	0	1	1	1	0
11	1	1	0	1	1	0	1	0	1	0	1
12	0	1	0	0	1	0	0	1	1	0	1
13	1	1	0	1	1	0	1	0	1	0	1
14	0	1	1	0	1	1	1	1	1	1	0

Table 9.6.

Test pattern no.	Test sequence			Sequences at circuit nodes							
	x_1	x_2	x_3	2	4	5	6	7	8	9	10
1	1	0	0	1	0	0	1	0	0	0	1
2	1	0	1	1	0	1	1	1	0	0	1
3	1	1	1	1	1	1	0	1	1	1	0
4	0	0	0	0	0	0	0	0	0	0	1
5	0	0	1	0	0	1	0	0	0	0	1
6	0	1	0	0	1	0	0	0	0	0	1
7	1	1	1	1	1	1	1	0	0	0	1
8	0	0	1	0	0	1	0	1	0	0	1
9	0	1	1	0	1	1	0	0	0	0	1
10	1	0	1	1	0	1	1	0	0	0	1
11	1	1	0	1	1	0	1	1	1	0	1
12	0	1	0	0	1	0	0	1	1	0	1
13	1	1	0	1	1	0	1	0	0	0	1
14	0	1	1	0	1	1	1	1	1	1	0

Table 9.7.

i	2	4	5	6	7	8	9	10
$S_i(x)$	U	1	9	0	1	A	U	P
$S_i^*(x)$	U	1	9	8	0	F	7	6

A comparison between signature $S_i^*(x)$ and their expected values shows that they differ at nodes 6, 7, 8, 9 and 10. In order to localize the fault that has occurred, the back-tracing technique, which does not allow one to decrease the number of faulty gates, is used first. Actually, the inequality $S_{10}^*(x) \neq S_{10}(x)$ initiates a test for correspondence between $S_9^*(x)$ and $S_9(x)$. Based on the difference between $S_9^*(x)$ and $S_9(x)$, the values $S_8^*(x)$ and $S_5^*(x)$ are tested. Since $S_8^*(x) \neq S_8(x)$, the values $S_7^*(x)$ and $S_4^*(x)$ are tested. When signature $S_7^*(x)$ does not match its expected value, signature $S_6^*(x)$, which should also be different from the expected value, is tested. Finally, the equality condition $S_2^*(x) = S_2(x)$ suggests that the fault is associated with the set of gates with a feedback loop placed around them. However, the fault on output NOT gate has not been detected. Then to test the hypothesis of a stuck-at fault to occur at the nodes of the circuit under test, the actual signatures are checked against the values $S(0) = 0$ and $S(1) = 1$ (see Table 9.4 for $l = 2^m - 2$). As a consequence, $S_4^*(x) = 1$ and $S_7^*(x) = 0$. To gain greater insight into why the sequence formed at node 7 is zero whereas that at node 4 is one, determine the values of signatures at the specified nodes for other l values. Thus suppose that for $l = 2^{4-1} - 1 = 7$, $S_4^*(x) = 0$ and $S_7^*(x) = A$, *which is contradictory to the hypothesis of sequences* $\{y(k)\} = 1$, $k = 1$ *to l, and* $\{y(k)\} = 0$, $k = 1$ *to l, to be generated at the nodes 4 and 7, respectively.* Thus we may conclude that the circuit under test is free of stuck-at faults.

Next the possibility for a bridging fault to occur is tested by making sure whether the condition $S_i^*(x) = S_j^*(x)$, $i \neq j \in \{2, 4, 5, 6, 7, 8, 9, 10\}$, is satisfied. It turns out that $S_i^*(x) \neq S_j^*(x)$ with $i \neq j$. Hence we may deduce that the circuit under test is free of bridging faults.

Further investigation of the circuit for fault localization is only possible by locating a fault in each of its gates individually. For that purpose, let us use corollaries to the following theorems.

Theorem 9.3. Signature $S_f(x)$ of a sequence generated at the output of an n-input EOR gate is determined as a bit-by-bit modulo-2 sum of signatures $S_i(x)$, $i = 1$ to n, for sequences produced at its inputs.

Proof. The input sequence produced at the ith input to the EOR gate, $i \in \{1, 2, \ldots, n\}$, is described by the polynomial $\kappa_i(x)$ whereas the sequence at its output is described by the sum

$$\kappa_f(x) = \sum_{i=1}^{n} \kappa_i(x) \tag{9.13}$$

as follows from the functional description of the EOR gate.

For every polynomial $\kappa_i(x)$, the relation (8.2)

$$\kappa_i(x) = q_i(x)\varphi^{-1}(x) \oplus S_i(x) \qquad i = 1 \text{ to } n$$

holds true. By substituting it into (9.13), we finally obtain

$$\kappa_f(x) = \varphi^{-1}(x) \sum_{i=1}^{n} q_i(x) \oplus \sum_{i=1}^{n} S_i(x).$$

Hence

$$S_j(x) = \overset{n}{\underset{i=1}{\sum}} S_i(x) \qquad (9.14)$$

as we wished to prove. ∎

Theorem 9.4. Signature $S_j(x)$ of an output sequence at the NOT gate is determined as a bit-by-bit modulo-2 sum of signatures $S_1(x)$ and $S(1)$, where $S_1(x)$ is the input sequence signature.

Proof. The logic function $f = \bar{x}$ may be represented as $f = x \oplus 1$ performed on a two-input EOR gate for which Theorem 9.3 holds true. Based on the theorem and by using (9.4) we obtain

$$S_j(x) = S_1(x) \oplus S(1) \qquad (9.15)$$

as we wished to prove. ∎

Thus by using Theorem 9.4 and the table of bit-by-bit sums of possible signatures (Table 9.8), we may determine that the NOT gate functions normally in the circuit of Fig. 9.1. From the table it follows that the modulo-2 sum of $S_9^*(x) = 7$ and $S(1) = 1$ is 6, which corresponds to the actual signature $S_{10}^*(x) = 6$ (see Table 9.7) and complies with the relation (9.15), which holds true for the fault-free NOT gate.

In a similar manner we may test whether the relation (9.14) holds true for a two-input EOR gate. As a consequence, we obtain that the modulo-2 sum $S_2^*(x) = U$ and $S_5^*(x) = 7$ is 8 and corresponds to the values $S_6^*(x)$. This implies that the relation (9.14) holds true and the hypothesis of the two-input EOR gate being fault-free is acceptable. To improve the confidence level of the hypothesis on normal behaviour

Table 9.8.

S(x)	0	1	2	3	4	5	6	7	8	9	A	C	F	H	P	U
0	0	1	2	3	4	5	6	7	8	9	C	A	F	H	P	U
1	1	0	3	2	5	4	7	6	9	8	C	A	H	F	U	P
2	2	3	0	1	6	7	4	5	A	C	8	9	P	U	F	H
3	3	2	1	0	7	6	5	4	C	A	9	8	U	P	H	F
4	4	5	6	7	0	1	2	3	F	H	P	U	8	9	A	C
5	5	4	7	6	1	0	3	2	H	F	U	P	9	8	C	A
6	6	7	4	5	2	3	0	1	P	U	F	H	A	C	8	9
7	7	6	5	4	3	2	1	0	U	P	H	F	C	A	9	8
8	8	9	A	C	F	H	P	U	0	1	2	3	4	5	6	7
9	9	8	C	A	H	F	U	P	1	0	3	2	5	4	7	6
A	A	C	8	9	P	U	F	H	2	3	0	1	6	7	4	5
C	C	A	9	8	U	P	H	F	3	2	1	0	7	6	5	4
F	F	H	P	U	8	9	A	C	4	5	6	7	0	1	2	3
H	H	F	U	P	9	8	C	A	5	4	7	6	1	0	3	2
P	P	U	F	H	A	C	8	9	6	7	4	5	2	3	0	1
U	U	P	H	F	C	A	9	8	7	6	5	4	3	2	1	0

of NOT and two-input EOR gates, the relations (9.14) and (9.15) are tested for other l values. When they hold true, the gates under test are considered to be fault-free. Thus the set of gates with the sought-for fault in the circuit under test will be defined only by two-input OR and AND gates and D-type flip-flop, with the possibility of indirectly checking whether the D-type flip-flop behaves normally by signatures obtained on its 1 and 0 inputs, direct and inverted sequences for which relation (9.12) is satisfied. However, it is not possible to state whether non-linear AND and OR gates behave normally by means of linear circuits used for signature generation. This stems from the major disadvantage of signature analysis, which lies in the fact that the actual signature differing from the expected one bears practically no additional information on whether the circuit under test behaves normally or not.

9.4. MULTI-FUNCTIONAL SIGNATURE ANALYSER

For more detail on the behaviour of a faulty digital circuit, it is reasonable to use a multi-functional signature analyser (Fig. 9.2), which allows one to determine some properties of the sequences produced on the nodes of the circuit under test and thus to estimate the serviceability of each gate individually.

Two modes of multi-functional analysis are possible depending on the value of control signal $V \in \{0, 1\}$. With $V = 1$, the analyser consists of two functionally independent devices. The first device is formed of a shift register (1) and a modulo-2 adder (4) representing a conventional serial signature analyser structure (see Fig. 8.1).

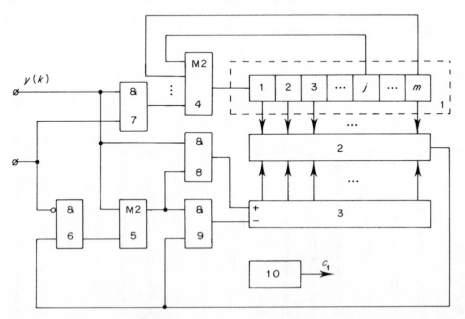

Fig. 9.2. Block diagram of a multi-functional signature analyser: 1, shift register; 2, comparison circuit (CC); 3, reversible counter; 4, modulo-2 adder; 5, two-input modulo-2 adder; 6, AND gate with inverting input; 7, 8, two-input AND gates; 9, three-input AND gate; 10, control unit

The value of signature $S(x) = a_1(l)a_2(l)\cdots a_m(l)$ is formed on its memory elements by the following linear relation (8.1):

$$a_i(0) = 0 \qquad i = 1 \text{ to } m$$

$$a_1(k) = y(k) \oplus \sum_{i=1}^{m} \alpha_i a_i(k-1)$$

$$a_j(k) = a_{j-1}(k-1) \qquad j = 2 \text{ to } m, \, k = 1 \text{ to } l$$

where $y(k) \in \{0, 1\}$ is the kth symbol of the sequence to be compressed, $\alpha_i \in \{0, 1\}$ is the ith coefficient of generating polynomial $\varphi(x)$, l is the length of tested sequence, which normally satisfies the condition $l \leq 2^m - 1$, and is specified by the control unit (10) as the number of clock pulses C_1.

The second device is formed of a reversible block counter (3), which is a totalizing meter counting the number $C(x)$ of 1s in the sequence $\{y(k)\}$, $k = 1$ to l, by the expression

$$C(x) = \sum_{k=1}^{l} y(k) \tag{9.16}$$

with $l \leq 2^m - 1$ where $m = \deg \varphi(x)$ and determines the width of counter 3.

When the control signal $V = 0$, the multi-functional signature analyser structure is transformed into that of a stochastic servo-integrator [72] consisting of a pseudorandom number generator (PRNG) employing units 1 and 4, a comparison circuit (CC) (2) and a reversible counter (RC) (3). The inputs to the latter are applied with the tested sequence $\{y(k)\}$, and the sequence from the CC output causes a 1 signal to be produced at the comparison circuit output if the binary code for a successive random number is less than the current code stored by RC. Otherwise, a 0 signal is produced.

As has been stated in [72], with $l > 3 \times 2^m$, the RC contents will indicate the probability $P(y) = P[y(k) = 1]$ for a 1 signal to appear in the sequence $\{y(k)\}$. Thus by using the multi-functional signature analyser (Fig. 9.2) for testing the sequence $\{y(k)\}$ we may obtain three major characteristics of the sequence: signature $S(x)$, number of 1s $C(x)$ in the sequence, and the probability $P(y)$ of their appearance. Let us evaluate these characteristics for some typical sequences $\{y(k)\}$ and determine their interrelation for NOT, EOR, OR and AND gates.

For a zero sequence $\{y(k)\} = 0$, $k = 1$ to l, the values of the associated characteristics take the following form:

$$S(x) = a_1(l)a_2(l)\cdots a_m(l) = 00\cdots 0$$
$$C(x) = 0$$
$$P(y) = 0. \tag{9.17}$$

At the same time, for the sequence $y(k) = 1$, $k = 1$ to l, we shall have

$$S(x) = a_1(l)a_2(l)\cdots a_m(l) = \begin{cases} 00\cdots 0 & \text{for } l = 2^m - 1 \\ 00\cdots 1 & \text{for } l = 2^m - 2 \end{cases} \tag{9.18}$$

$$C(x) = l \qquad l \leq 2^m - 1$$
$$P(y) = 1 \qquad l > 3 \times 2^m.$$

To define the value $S(x)$ for the sequence $y(k) = 2]k/2[- k$, $k = 1$ to l, where $]k/2[= \text{int}(k/2)$ is the nearest integer that is greater than or equal to $k/2$, let us examine the following theorem.

Theorem 9.5. The value of signature $S_1(x)$ for the sequence of symbols $y_1(k) = 2]k/2[- k$, $k = 1$ to $2^m - 1$, corresponds to that for its reciprocal and is determined by the relation $S_1(x) = a_1^1(2^m - 1)a_2^1(2^m - 1)\cdots a_m^1(2^m - 1) = 11\cdots 1$.

Proof. Suppose that signature $S_1(x) = a_1^1(2^m - 1)a_2^1(2^m - 1)\cdots a_m^1(2^m - 1)$ corresponds to the sequence of symbols $y_1(k) = 2]k/2[- k$, $k = 1$ to $2^m - 1$, and signature $S_2(x) = a_1^2(2^m - 1)a_2^2(2^m - 1)\cdots a_m^2(2^m - 1)$ corresponds to its reciprocal. The values of these signatures are identical since the summed sequence $\{y(k)\} = \{y_1(k)\} \oplus \{y_2(k)\}$, $k = 1$ to $2^m - 1$, is a unit sequence $\{y(k)\}$, where $y(k) = 1$, $k = 1$ to $2^m - 1$, which has the zero signature $S(x) = 00\cdots 0$; hence, $S_1(x) \oplus S_2(x) = S(x) = 00\cdots 0$ and

$$a_i^1(2^m - 1) = a_i^2(2^m - 1) \qquad i = 1 \text{ to } m. \qquad (9.19)$$

Considering that the first zero symbol of sequence $\{y_2(k)\}$ does not affect the final value of signature $S_2(x)$, we may deduce that $\{y_2(k)\}$ compressed by the signature analyser produces a result that is equivalent to compressing the first $2^m - 2$ symbols of $\{y_1(k)\}$. Therefore, the values of symbols $a_i^2(2^m - 1) \in \{0, 1\}$, $i = 1$ to m, in signature $S_2(x)$ may be used for determining $S_1(x)$ by (8.1). Then, subject to

$$a_i^1(2^m - 2) = a_i^2(2^m - 1)$$

we obtain

$$a_1^1(2^m - 1) = y_1(2^m - 1) \oplus \sum_{i=1}^{m} \alpha_i a_i^2(2^m - 1)$$

$$a_j^1(2^m - 1) = a_{j-1}^2(2^m - 1) \qquad j = 2 \text{ to } m. \qquad (9.20)$$

Substituting the obtained values $a_i^1(2^m - 1)$ into the set of equations (9.19) and subjected to $y_1(2^m - 1) = 2]2^{m-1} - 2^{-1}[- 2^m + 1 = 2^m - 2^m + 1 = 1$, we shall finally have

$$a_i^2(2^m - 1) = 1 \oplus \sum_{i=1}^{m} \alpha_i a_i^2(2^m - 1)$$

$$a_1^2(2^m - 1) = a_2^2(2^m - 1) = a_3^2(2^m - 1) = \cdots = a_m^2(2^m - 1). \qquad (9.21)$$

An evident solution of the system of equations (9.21) is that all its unknowns $a_i^2(2^m - 1)$, $i = 1$ to m, be one. Thus, $S_1(x) = 11\cdots 1$, which proves the statement. ∎

Based on Theorem 9.5, the sequence of the form $\{y(k)\} = 101010\cdots 01$ will have the following integral characteristics:

$$S(x) = a_1(l)a_2(l)\cdots a_m(l) = 11\cdots 1 \qquad \text{for } l = 2^m - 1 \text{ and } l = 2^m - 2$$
$$C(x) = 2^{m-1} \qquad \text{for } l = 2^m - 1 \qquad\qquad (9.22)$$
$$P(y) = 1/2 \qquad \text{for } l > 3 \times 2^m.$$

The above characteristics can be similarly defined for more complex forms of sequence $\{y(k)\}$.

By using the results that have been obtained, we may establish the interrelation of characteristics $S(x)$, $C(x)$ and $P(y)$ for some logic elements. We shall represent the signatures in the unitary code as before, whereas the values $C(x)$ and $P(y)$ will be represented in the positional code. Let us consider first the NOT gate, for which the relations

$$S_j(x) = S_1(x) \oplus S(1)$$
$$C_j(x) = l - C_1(x) \qquad\qquad (9.23)$$
$$P_j(y) = 1 - P_1(y) \qquad l > 3 \times 2^m$$

are satisfied, where $S_j(x)$, $C_j(x)$ and $P_j(y)$ are the characteristics of the sequence of length l formed at the output of the NOT gate that realizes the function $f = \bar{x}$; $S_1(x)$, $C_1(x)$ and $P_1(y)$ are the respective characteristics of its input sequence. The value $S(1)$ represents the signature of a sequence that consists of l 1s.

For a multi-input EOR gate, subject to (9.14), we may obtain

$$S_j(x) = \overset{m}{\underset{i=1}{\bigoplus}} S_i(x). \qquad\qquad (9.24)$$

The expressions for $C_j(x)$ and $P_f(y)$ bear very complex functional dependence on the values of $C_i(x)$ and $P_i(y)$, $i = 1$ to n, of the n-input EOR gate and therefore the calculations of $C_j(x)$ and $P_j(y)$ are rather cumbersome. Thus, for $n = 2$, $P_j(y)$ has the form

$$P_j(y) = P_1(y) + P_2(y) - 2P_1(y)P_2(y).$$

Therefore, to check whether the EOR gate behaves normally, its signatures are usually analysed by the expression (9.24).

At the same time, the behaviour of AND and OR gates cannot be tested by their signatures $S(x)$, as has been shown in Sec. 9.3, although the relations for checking the specified gate behaviour may be obtained for $C(x)$ and $P(y)$. Thus for an OR gate, we may show that the value $C_f(x)$ satisfies the relation

$$\max\{C_1(x), C_2(x), ..., C_n(x)\} \leqslant C_f(x) \leqslant Q_1 \qquad\qquad (9.25)$$

where

$$Q_1 = \begin{cases} C_1(x) + C_2(x) + \cdots + C_n(x) & \text{for } C_1(x) + C_2(x) + \cdots + C_n(x) < l \\ l & \text{for } C_1(x) + C_2(x) + \cdots + C_n(x) \geqslant l \end{cases}$$

and n is the number of inputs to the OR gate.

The relation (9.25) holds true for a random functional dependence of input sequences in the gate of interest. Under the assumptions made, the following inequality is satisfied for the same gate similar to (9.25):

$$\max\{P_1(y), P_2(y), ..., P_n(y)\} \leqslant P_f(y) \leqslant Q_2 \qquad\qquad (9.26)$$

where

$$Q_2 = \begin{cases} P_1(y) + P_2(y) + \cdots + P_n(y) & \text{for } P_1(y) + P_2(y) + \cdots + P_n(y) < 1 \\ 1 & \text{for } P_1(y) + P_2(y) + \cdots + P_n(y) \geqslant 1. \end{cases}$$

At the same time, the relation

$$P_f(y) = 1 - \prod_{i=1}^{m} [1 - P_i(y)]$$

which assumes the form

$$P_f(y) = P_1(y) + P_2(y) - P_1(y)P_2(y)$$

with $n = 2$, holds true for $P_f(y)$ when the symbols of the input sequences of an OR gate are independent (see Sec. 5.1).

Moreover, for input sequences whose unit symbols are non-concurrent events, the equality $P_f(y) = Q_2$ is satisfied.

Relation (9.25), the same as (9.26), holds true for random input sequences although in this case the value of $C_f(x)$ can be estimated to a higher accuracy, provided that there is additional information on test stimuli, for example. This may be illustrated by the implementation of syndrome testing (see Sec. 4.3).

For an AND gate that realizes the function $f = x_1 x_2 \cdots x_n$, the following interrelation holds true for integral characteristics $C(x)$ of its input and output sequences:

$$Q_3 \leqslant C_f(x) \leqslant \min \{C_1(x), C_2(x), ..., C_n(x)\} \tag{9.27}$$

where

$$Q_3 = \begin{cases} 0 & \text{for } C_1(x) + C_2(x) + \cdots + C_n(x) \leqslant (n-1)l \\ nl - C_1(x) - C_2(x) - \cdots - C_n(x) & \text{for } C_1(x) + C_2(x) + \cdots + C_n(x) > (n-1)l. \end{cases}$$

Similarly to relation (9.27), for the gate in question the inequality

$$Q_4 \leqslant P_f(y) \leqslant \min \{P_1(y), P_2(y), ..., P_n(y)\} \tag{9.28}$$

where

$$Q_4 = \begin{cases} 0 & \text{for } P_1(y) + P_2(y) + \cdots + P_n(y) \leqslant n-1 \\ n - P_1(y) - P_2(y) - \cdots - P_n(y) & \text{for } P_1(y) + P_2(y) + \cdots + P_n(y) > n-1 \end{cases}$$

is satisfied.

Thus, the expressions obtained for the integral characteristics of sequences produced on the nodes of NOT, EOR, AND and OR gates allow one to check their behaviour.

When the above gates normally, the experimental characteristics must satisfy the relations (9.23)–(9.28). If the above relations are not satisfied for any of the characteristics being tested, we may infer that the gate under test is faulty. As an example of using the obtained expressions for fault detection in a circuit, let us check whether the OR and AND gates behave normally in the circuit, Fig. 9.1. The sequences summarized in Table 9.6 are generated on the nodes of the circuit. In accordance with their actual form in the circuit under test, calculate the values of $C(x)$ and $P(y)$. Thus for the sequence $y_4(k) = 00100110101111$ generated at the fourth node of the circuit, the specified characteristics will take the form $C_4(x) = 8$, $P_4(y) = 8/14 = 0.571$.

Table 9.9.

i	2	4	5	6	7	8	9	10
$C_i(x)$	7	8	8	7	6	4	2	12
$P_i(y)$	0.5	0.57	0.57	0.5	0.43	0.29	0.14	0.86

The complete set of values $C(x)$ and $P(y)$ for the circuit under test (see Fig. 9.1) is given in Table 9.9.

Using the obtained characteristics of the circuit under test, estimate the behaviour of its two-input AND gate for which the inequalities (9.27) and (9.28) must be satisfied. The values Q_3 and Q_4 in this case will assume the values $Q_3 = 0$, since $C_5(x) + C_8(x) = 8 + 4 < 14$, and Q_4, since $P_5(y) + P_8(y) = 0.57 + 0.29 < 1$. Then substituting the obtained values into relations (9.27) and (9.28), we obtain

$$0 \leqslant C_9(x) = 2 \leqslant C_8(x) = 4 \qquad \text{and} \qquad 0 \leqslant P_9(y) = 0.14 \leqslant P_8(y) = 0.29$$

which indicate that the gate under test is fault-free. However, the values $C_7(x) = 6$, $C_4(x) = 8$ and $C_8(x) = 4$ obtained for the OR gate do not satisfy the inequality (9.25), which in this case assumes the form

$$\max \{C_4(x), C_7(x)\} \leqslant C_8(x) \leqslant C_4(x) + C_7(x).$$

In the same way, the values $P_4(y) = 0.57$, $P_7(y) = 0.43$ and $P_8(y) = 0.29$ do not satisfy the inequality (9.26)

$$\max \{P_4(y), P_7(y)\} \leqslant P_8(y) \leqslant 1.$$

Thus from the fact that relations (9.25) and (9.26) are not satisfied for the OR gate, we may deduce that this gate in the circuit behaves abnormally and the sought fault of the circuit of Fig. 9.1 is associated with the faulty OR gate.

Further improvement of signature analyser efficiency and resolution expressed as the fault localization capability may be attained by modifying the basic design of signature analyser as well as by applying the new techniques for compressing the sequences under test.

10

NEW SIGNATURE GENERATION TECHNIQUES FOR BINARY SEQUENCES

10.1. OUTPUT RESPONSE COMPRESSION BY SHIFT REGISTERS

The idea behind compressing output responses of a digital circuit by shift registers [134, 135] is in many ways similar to the basic principles of signature analysis. Such a circuit configured of a shift register and a modulo-2 adder is described by polynomial $\varphi(x) = 1 \oplus x^m$, where $m = \deg \varphi(x)$ is the number of memory elements in the shift register. The compression circuit based on a shift register functions in accordance with a system of logic equations that have been obtained by (8.1) for a polynomial $\varphi(x) = 1 \oplus x^m$:

$$a_1(k) = y(k) \oplus a_m(k-1)$$
$$a_j(k) = a_{j-1}(k-1) \qquad j = 2 \text{ to } m, \ k = 1, 2, 3, \ldots \qquad (10.1)$$

where $a_j(k) \in \{0, 1\}$ is the contents of the jth memory element of the shift register in the kth cycle of its operation; $y(k) \in \{0, 1\}$ is the value of a binary character applied to the compression circuit input in the kth cycle. Without loss of generality, suppose that $k = 1$ to l, where $l = 2^m - 1$. Then by successively applying the set of equations (10.1) for compression of output response $y(1)$, $y(2)$, $y(3), \ldots, y(l)$ we shall obtain

$$a_i(l) = \sum_{n=0}^{\lambda} y(l - nm - i + 1) \oplus a_{(m-l_0+i-1) \bmod m}(0) \qquad i = 1 \text{ to } l_0 + 1$$

$$a_i(l) = \sum_{n=0}^{\lambda-1} y(l - nm - i + 1) \oplus a_{(m-l_0+i-1) \bmod m}(0) \qquad i = l_0 + 2 \text{ to } m \quad (10.2)$$

in l cycles where $l_0 = l \bmod m$, $\lambda = (l - l_0)/m$.

For $m = 5$, we obtain $l = 2^5 - 1 = 31$, $l_0 = 1$ and $\lambda = 6$. Then the set of equations (10.2) takes on the form

$$a_1(31) = y(31) \oplus y(26) \oplus y(21) \oplus y(16) \oplus y(11) \oplus y(6) \oplus y(1) \oplus a_4(0)$$
$$a_2(31) = y(30) \oplus y(25) \oplus y(20) \oplus y(15) \oplus y(10) \oplus y(5) \oplus y(0) \oplus a_5(0)$$

$$a_3(31) = y(29) \oplus y(24) \oplus y(19) \oplus y(14) \oplus y(9) \oplus y(4) \oplus a_1(0)$$
$$a_4(31) = y(28) \oplus y(23) \oplus y(18) \oplus y(13) \oplus y(8) \oplus y(3) \oplus a_2(0)$$
$$a_5(31) = y(27) \oplus y(22) \oplus y(17) \oplus y(12) \oplus y(7) \oplus y(2) \oplus a_3(0).$$

In the general case, the initial state of the shift register may assume a random value, thereby causing no effect on the result of compressing the output response of the digital circuit. To simplify the discussion that follows, suppose that $a_1(0) = a_2(0) = a_3(0) = \cdots = a_m(0) = 0$; then the set of equations (10.2) will take the form

$$a_i(l) = \sum_{n=0}^{\lambda} y(l - nm - i + 1) \qquad i = 1 \text{ to } l_0 + 1$$

$$a_i(l) = \sum_{n=0}^{\lambda-1} y(l - nm - i + 1) \qquad i = l_0 + 2 \text{ to } m. \tag{10.3}$$

Thus the contents of the shift register in l cycles of its operation will be the signature of sequence $y(1), y(2), y(3), \ldots, y(l)$,

$$S(y) = a_1(l)a_2(l)a_3(l) \cdots a_m(l). \tag{10.4}$$

The reported value of signature $S(y)$ is the reference one for sequence $\{y(k)\}$, which may contain error values in the general case. Any error sequence may be described by an error polynomial or an error sequence $\{z(k)\}$, which has the same length as the sequence $\{y(k)\}$ and whose symbol $z(k)$, $k = 1$ to l, is zero if the actual value of $y(k)$ is true. Otherwise it will be one. Thus the actual sequence $\{y*(k)\}$ may be represented as a bit-by-bit sum modulo-2 of sequences $\{y(k)\}$ and $\{z(k)\}$ and the signature $S*(y)$ equals $S(y) \oplus S(z)$. When the sequence $\{y(k)\}$ is error-free, $S(z) = 000 \cdots 0$ and hence $S*(y) = S(y)$ [135]; when the sequence has some errors, $S(z)$ may assume an arbitrary value and thus with $S(z) = 000 \cdots 0$ the sequence of error bits described by $\{z(k)\}$ will be undetectable. In the remaining cases, the faults characterized by sequences $\{z(k)\}$ for which $S(z) \neq 000 \cdots 0$ are detectable since

$$S*(y) = S(y) \oplus S(z) \neq S(y).$$

This implies that the real signature $S*(y)$ does not correspond to the reference $S(y)$.

Let us measure the efficiency of the discussed technique for output response compression. We prove the following theorem first [134].

Theorem 10.1. For the ith symbol ($i = 1$ to m) of signature $S(z) = a_1(l)a_2(l) \cdots a_m(l)$ of a random sequence $\{z(k)\}$ for which $0 < P[z(k) = 1] < 1$, $k = 1$ to l, the equality

$$\lim_{l \to \infty} P[a_i(l) = 1] = 0.5$$

holds true.

Proof. Consider first the value $a_i(k)$ for $i = 1$, which is determined by (10.1) as $z(k) \oplus a_1(k - m)$. Then

$$P[a_1(k) = 1] = P[z(k) \oplus a_1(k - m) = 1].$$

From the relation $z(k) \oplus a_1(k - m) = \overline{z(k)}a_1(k - m) + z(k)\overline{a_1(k - m)}$ and independence of events $z(k)$ and $z(k-1)$, we may obtain

$$P[a_1(k) = 1] = \{1 - P[z(k) = 1]\}P[a_1(k - m) = 1]$$
$$+ P[z(k) = 1]\{1 - P[a_1(k - m) = 1]\}$$
$$= P(a_1(k - m) = 1] + P[z(k) = 1]\{1 - 2P[a_1(k - m) = 1]\}. \quad (10.5)$$

Assuming that the probability $P[a_1(k - m) = 1]$ differs from 0.5 by δ_{k-m}, define δ_k for the probability $P[a_1(k) = 1]$. By substituting $P[a_1(k - m) = 1] = 0.5 + \delta_{k-m}$ and $P[a_1(k) = 1] = 0.5 + \delta_k$ into (10.5), we obtain

$$\delta_k = \delta_{k-m} - 2P[z(k) = 1]\delta_{k-m}.$$

Hence, subject to inequality $P[z(k) = 1] < 1$, we obtain that $|\delta_k| < |\delta_{k-m}|$. Therefore $\lim_{k \to \infty} \delta_k = 0$, for formally substituting k for l, we obtain

$$\lim_{l \to \infty} P[(a_1(l) = 1] = 0.5. \quad (10.6)$$

The relation (10.6) also holds true for $a_2(l)$, $a_3(l), \ldots, a_m(l)$ since by (10.1) $a_j(k) = a_1(k + 1 - j)$ and for $a_1(k + 1 - j)$ the relation in question holds true, as was to be shown. ∎

By using the corollary to the previous theorem, let us formulate and prove the following statement [134].

Theorem 10.2. For the reference signature $S(y)$ and a signature $S^*(y)$ of random sequence $\{y^*(k)\}$, $k = 1$ to l, the limiting relation

$$\lim_{l \to \infty} P[S^*(y) = S(y)] = 1/2^m$$

where l is the number of symbols in sequences $\{y(k)\}$ and $\{y^*(k)\}$, m is the length of signatures $S(y)$ and $S^*(y)$, holds true.

Proof. By Theorem 10.1, each symbol of signatures $S^*(y)$ of a random sequence $\{y^*(k)\}$ may equally likely assume a 0 or 1 for $l \to \infty$. Thus, considering (10.3), we may demonstrate that the specific value assumed by any symbol of $S^*(y)$ is independent. Therefore, the probability of $S^*(y) = S(y)$ may be determined as

$$P[S^*(y) = S(y)] = P[a_1^*(l) = a_1(l)]P[a_2^*(l) = a_2(l)] \times \cdots \times P[a_m^*(l) = a_m(l)]$$

where $a_i(l) \in \{0, 1\}$, $i = 1$ to m. Hence by (10.6) we finally obtain

$$\lim_{l \to \infty} P[S^*(y) = S(y)] = 1/2^m$$

as was to be shown. ∎

Theorem 10.2 allows one to measure the efficiency of the output response compression technique for digital circuits employing shift registers. By using the integral characteristic represented as probability P_n of failing to detect an error in the random sequence $\{y^*(k)\}$ as a measure, we can define by Theorem 10.2

$$P_n = P[S^*(y) = S(y)] \simeq 1/2^m. \quad (10.7)$$

The relation (10.7) obtained for fairly general limitations on the form of sequence to be compressed does not allow one to compare the discussed compression technique against other popular techniques and, in particular, with the well studied signature analysis technique. Therefore, we shall consider some features of the described technique as compared with signature analysis.

(i) The shift-register compression technique allows one to detect all odd errors in the sequence $\{y^*(k)\}$, with a single error causing a change in only one bit of reference signature $S(y)$. A triple error may cause a change in one, two or three bits of $S(y)$, etc. Therefore, when μ bits in $S^*(y)$ do not correspond to reference $S(y)$, we may conclude that an error whose multiplicity is greater than or equal to μ has occurred in the sequence $\{y^*(k)\}$. Note that signature analysis allows one to detect only all single errors from the set of odd errors, and the form of the actual signature does not allow one to estimate all possible errors that occur in $\{y^*(k)\}$.

(ii) For an equiprobable configuration of error bits in the sequence $\{y^*(k)\}$, let us estimate the probability P_n^μ of failing to detect an error whose $\mu = 2$ for the shift-register compression technique. The number Q_n^μ of undetectable errors for $\mu = 2$ will be determined by the conditions when both error bits participate in generation of the same ith bit of signature $S^*(y)$. Then $a_i^*(l) = a_i(l)$ and, hence, $S^*(y) = S(y)$, implying that the mentioned error is undetectable. The value of Q_n^2 is

$$Q_n^2 = l_0 C_{\lambda+1}^2 + (m - l_0)C_\lambda^2$$

and the probability P_n^2

$$P_n^2 = \frac{l_0 C_{\lambda+1}^2 + (m - l_0)C_\lambda^2}{C_{m\lambda + l_0}^2}$$

where $C_{m\lambda + l_0}^2$ is the total number of possible double errors in the sequence of length $l = m\lambda + l_0$. For $l = 15$ and $m = 4$, we obtain that

$$P_n^2 = \frac{3C_4^2 + (4 - 3)C_3^2}{C_{15}^2} = \frac{7}{35}.$$

We can similarly find the value P_n^μ for errors of random multiplicity μ whose analysis reveals their non-uniform distribution P_n^μ. Therefore, when measuring the efficiency of the shift-register compression technique, it is necessary to know probability distribution P_v^μ of errors with multiplicity μ in each specific case. Only by comparative analysis of probability P^μ of failing to detect the error of multiplicity μ, defined by (8.17), may we establish whether the technique of interest is suitable.

The most often proposed technique is used when $P_v^\mu = 0$ for all even values of μ, hence $P^\mu = 0$. This implies that all error sequences $\{y^*(k)\}$ are detectable.

The shift-register compression technique has been extended to the analysis of multi-output circuit responses [136]. The techniques proposed may be used for designing multi-line structures which handle several input symbols in each cycle.

A characteristic property of the shift-register compression technique is the source sequence $\{y(k)\}$ partitioning into multiple subsequences by which the symbols of signature $S(y)$ are produced independently. For implementing the so-called spectral data compression technique, the $\{y(k)\}$ partitioning algorithms are more complex.

10.2. SPECTRAL TECHNIQUE FOR ESTIMATING OUTPUT RESPONSES OF DIGITAL CIRCUITS

The technique of interest is a further development of syndrome testing [59] based on generation of the exhaustive test pattern sequence and determination of 1s produced at the output of the circuit that implements a Boolean function F. Then the syndrome $S(F)$ of F corresponds to coefficient r_0 of the zero power of the Walsh spectrum for the specified function.

The use of two Walsh coefficients for testing is dealt with in [137]. Further investigations on the use of the Walsh coefficients for estimating the state of a circuit under test have been reported in [138, 139]. Consider the basic results reported there. For this purpose, we shall introduce the notion of the Walsh coefficient for the Boolean function F.

By terminology adopted in [138, 139], the Walsh coefficients of F are calculated as

$$r = T_n F \tag{10.8}$$

where T_n is the $2^n \times 2^n$ matrix

$$T_n = \begin{vmatrix} T_{n-1} & T_{n-1} \\ T_{n-1} & -T_{n-1} \end{vmatrix} \qquad T_0 = |1| \tag{10.9}$$

are r and F are respectively the Walsh coefficient vector and the vector of values for $F = F(x_1, \ldots, x_n)$ which correspond to its truth table.

For the case of a Boolean function of three variables $F(x_1, x_2, x_3) = x_1 \bar{x}_3 + \bar{x}_1 \bar{x}_2 x_3$ described by the truth table (Table 10.1) [138], we obtain by using the matrix T_3 based on (10.9) that the Walsh coefficients r will be calculated as

$$
\begin{vmatrix} r_0 \\ r_1 \\ r_2 \\ r_{12} \\ r_3 \\ r_{13} \\ r_{23} \\ r_{123} \end{vmatrix}
=
\begin{vmatrix}
1 & 1 & 1 & 1 & 1 & 1 & 1 & 1 \\
1 & -1 & 1 & -1 & 1 & -1 & 1 & -1 \\
1 & 1 & -1 & -1 & 1 & 1 & -1 & -1 \\
1 & -1 & -1 & 1 & 1 & -1 & -1 & 1 \\
1 & 1 & 1 & 1 & -1 & -1 & -1 & -1 \\
1 & -1 & 1 & -1 & -1 & 1 & -1 & 1 \\
1 & 1 & -1 & -1 & -1 & -1 & 1 & 1 \\
1 & -1 & -1 & 1 & -1 & 1 & 1 & -1
\end{vmatrix}
\times
\begin{vmatrix} 0 \\ 1 \\ 0 \\ 1 \\ 1 \\ 0 \\ 0 \\ 0 \end{vmatrix}
=
\begin{vmatrix} 3 \\ -1 \\ 1 \\ 1 \\ 1 \\ -3 \\ -1 \\ -1 \end{vmatrix}
$$

Table 10.1.

x_3	x_2	x_1	$F(x_1, x_2, x_3)$
0	0	0	0
0	0	1	1
0	1	0	0
0	1	1	1
1	0	0	1
1	0	1	0
1	1	0	0
1	1	1	0

where, in particular, the coefficient $r_0 = 3$ equals the syndrome $S(F)$ of the specified function multiplied by 2^3. In the general form the analytic expression for the Walsh coefficient r_0 may be written as

$$r_0 = \sum_{x_1 \cdots x_n = 0 \cdots 0}^{1 \cdots 1} F(x_1, \ldots, x_n).$$

The coefficient $r_i, i = 1$ to n, is a measure of the correlation between $F(x_1, \ldots, x_n)$ and the variable x_i,

$$r_i = \sum_{x_1 \cdots x_n = 0 \cdots 0}^{1 \cdots 1} [(1 - 2x_i)F(x_1, \ldots, x_n)] \qquad (10.10)$$

where characters $\sum$ and $-$ denote the arithmetic addition and subtraction operations.

The high-order coefficients measure the correlation between $F(x_1, \ldots, x_n)$ and the respective modulo-2 sum of its variables. Thus, for example, we may find r_{123} by the relation

$$r_{123} = \sum_{x_1 \cdots x_n = 0 \cdots 0}^{1 \cdots 1} \{[1 - 2(x_1 \oplus x_2 \oplus x_3)]F(x_1, \ldots, x_n)\}.$$

The introduced concept of Walsh coefficients allows one to define syndrome testing of a digital circuit in terms of the Walsh coefficients of the Boolean function $F = F(x_1, \ldots, x_n)$ implemented by the circuit.

As in Theorem 4.5, the following statement also holds true.

Theorem 10.3. The two-level non-redundant circuit described by a Boolean function $F = F(x_1, \ldots, x_n)$ is syndrome-testable with respect to an input x_i if and only if the inequality $r_i \neq 0$, $i \in \{1, 2, 3, \ldots, n\}$, is satisfied.

Proof. The original function $F = F(x_1, \ldots, x_n)$ may be represented as $F = G_1 x_i + G_2 \bar{x}_i + G_3, i \in \{1, 2, 3, \ldots, n\}$, where $G_1 \neq 0$, $G_2 \neq 0$, $G_3 \neq 1$, $\partial G_1/\partial x_i = \partial G_2/\partial x_i = \partial G_3/\partial x_i = 0$. Then by (10.10) we obtain

$$r_i = \sum_{x_1 \cdots x_n = 0 \cdots 0}^{1 \cdots 1} [(1 - 2x_i)(G_1 x_i + G_2 \bar{x}_i + G_3)].$$

The latter expression may be represented as two terms

$$r_i = \sum_{x_1 \cdots x_i - 1 x_i + 1 \cdots x_n = 0 \cdots 0}^{1 \cdots 1} (G_2 + G_3) - \sum_{x_1 \cdots x_i - 1 x_i + 1 \cdots x_n = 0 \cdots 0}^{1 \cdots 1} (G_1 + G_3)$$

each of which has the form of syndromes $S(G_2 + G_3)$ and $S(G_1 + G_3)$ of functions $G_2 + G_3$ and $G_1 + G_3$ multiplied by the coefficient 2^n. When the inequality $r_i \neq 0$ is satisfied, the syndrome $S(G_2 + G_3)$ differs from $S(G_1 + G_3)$, i.e. $S(G_2 + G_3) \neq S(G_1 + G_3)$; hence by (4.9) and (4.11), $S(G_2 \bar{G}_3) \neq S(G_1 \bar{G}_3)$, which, by theorem 4.5, is in its turn the condition for circuit testability with respect to x_i as a result of syndrome analysis. Thus, with $r_i \neq 0$, the digital circuit is syndrome-testable, as was to be shown. ■

When the equality $r_i = 0$ is satisfied, the digital circuit is syndrome-untestable with respect to input x_i. For the above discussed example of Boolean function

$F(x_1, x_2, x_3) = x_1\bar{x}_3 + \bar{x}_1\bar{x}_2 x_3$, the circuit that implements the function will be syndrome-testable with respect to input nodes x_1, x_2 and x_3 since the Walsh coefficients r_1, r_2 and r_3 are non-zero and assume the values $r_1 = -1, r_2 = 1$ and $r_3 = 1$.

More general results that allow one to estimate syndrome testing of internal nodes of a digital circuit follow from the following theorem, presented here without proof [138, 139].

Theorem 10.4. In a digital circuit that implements $F = F(x_1, \ldots, x_n) = G[x_1, \ldots, x_n, g(x_1, \ldots, x_n)]$ the inner node described by $g(x_1, \ldots, x_n)$ is syndrome-testable with respect to fault $g \equiv 0$ since

$$r_0 \neq \tfrac{1}{2}(r_0^* + r_{n+1}^*)$$

or with respect to $g \equiv 1$ since

$$r_0 \neq \tfrac{1}{2}(r_0^* - r_{n+1}^*)$$

where r_0 is the Walsh spectrum coefficient of F; r_0^* and r_{n+1}^* are respectively the Walsh coefficients for function G.

Theorems 10.3 and 10.4 may be applied for analysing the testability of circuit nodes by a single Walsh coefficient r_0, which corresponds to the syndrome of the function implemented by the circuit. For the case when the zero-order Walsh coefficient r_0 does not allow one to detect a fault, the high-order coefficients may be used for the purpose as reported in [138]. Then the resulting signature will contain several Walsh coefficients.

The concept of syndrome testing of a digital circuit is a particular case of r_α-testability, which is determined as follows.

A digital circuit is considered to be r_α-testable with respect to the specified fault only when the Walsh coefficient r_α of the faulty circuit differs from its reference value. The value of α may be defined from the set $\{0, 1, \ldots, n\}$. With $\alpha = 0$, r_0-testability will correspond to the syndrome-testability of the circuit.

Theorems 10.3 and 10.4 hold true for the case of r_0-testability and may be generalized to oother cases of r_0-testability. Let us state two theorems for r_α-testability without proof [139].

Theorem 10.5. A two-level non-redundant digital circuit described by a Boolean function $F = F(x_1, \ldots, x_n)$ is r_α-testable with respect to the input x_i if and only if $r_{\alpha i} \neq 0$, where $\alpha, i \in \{1, 2, \ldots, n\}, \alpha \neq i$.

Theorem 10.6. An inner node of a digital circuit implementing $F = F(x_1, \ldots, x_n) = G[x_1, \ldots, x_n, g(x_1, \ldots, x_n)]$ that is described by $g(x_1, \ldots, x_n)$ is r_α-testable if the following system of inequalities

$$r_\alpha \neq \tfrac{1}{2}[r_\alpha^* + r_{\alpha(n+1)}^*]$$

$$r_\alpha \neq \tfrac{1}{2}[r_\alpha^* - r_{\alpha(n+1)}^*]$$

where r_α, r_α^* and $r_{\alpha(n+1)}^*$ are the Walsh spectrum coefficients for Boolean functions F and G, hold true.

An example of applicability of the r_α-testability concept may be a digital circuit that implements the Boolean function $F(x_1, x_2, x_3) = x_2 + x_1x_3 + \bar{x}_1\bar{x}_3$. The Walsh spectrum for this function will be represented by its coefficients $r_0 = 6, r_1 = 0, r_2 = -2$, $r_{12} = 0, r_3 = 0, r_{13} = 2, r_{23} = 0, r_{123} = 0$ obtained by (10.8).

The analysis of Walsh coefficients shows that the input nodes x_1 and x_3 in the circuit that implements $F(x_1, x_2, x_3)$ are r_0-untestable. This means that their possible fault will not be detected by analysing the actual syndrome $S(F)$, since $r_1 = r_3 = 0$ (see Theorem 10.3). At the same time, $r_{13} \neq 0$, thereby implying that x_1 is r_3-testable and x_3 is r_1-testable. Thus a signature that might detect the faults on nodes x_1, x_2 and x_3 in the circuit that implements $F(x_1, x_2, x_3)$ consists of the values of three Walsh spectrum coefficients for F, i.e. r_0, r_1 and r_3 [138].

Another example of r_α-testability is the circuit of Fig. 10.1. Consider its r_0-testability with respect to input nodes x_1, x_2, x_3, x_4 and the output of the OR gate [138].

The Walsh spectrum coefficients for $F(x_1, x_2, x_3, x_4)$, namely r_0, r_1, r_2, r_3 and r_4, assume the values 6, $-2, -2, -2$ and -2. Hence the inner nodes in the circuit (Fig. 10.1) are r_0-testable. At the same time, for the inner node of circuit f_1 we obtain $r_5^* = 0$ and $r_0^* = 12$, hence

$$r_0 = \tfrac{1}{2}(r_0^* + r_5^*) = \tfrac{1}{2}(r_0^* - r_5^*).$$

This implies that the faults are syndrome-untestable (r_0-untestable) for the node. However, for coefficients $r_4^* = 4, r_{45}^* = -4$, the inequality

$$r_4 \neq \tfrac{1}{2}(r_4^* \pm r_{45}^*)$$

is satisfied, which allows one to conclude that the faults are r_4-testable for the inner node f_1. Thus a signature that allows one to detect faults on nodes x_1, x_2, x_3, x_4 and f_1 is the value of two coefficients r_0 and r_4 from the Walsh spectrum of $F(x_1, x_2, x_3, x_4)$.

The examples discussed above show that, in the general case, more than one Walsh spectrum coefficient is required to ensure complete testability of a digital circuit. The $n + 1$ coefficient is used as the upper bound of coefficients required for the n-input combinational circuit [138].

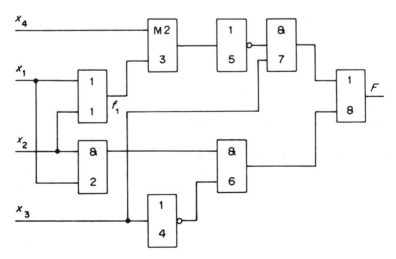

Fig. 10.1. A digital circuit

A major problem encountered on using the spectral technique for digital circuit testing is an analysis for r_α-testability. Furthermore, the actual signatures have rather high dimension determined by the number of Walsh coefficients that have been used. The complexity of Walsh coefficient calculations is also worth mentioning. The scheme for their implementation comprises a binary counter producing all possible input combinations, a Walsh function generation, a circuit under test and an estimator for the Walsh spectrum coefficients [138]. In some cases, fast Walsh transforms may be used for estimating r_α-testability.

10.3. CORRELATION TECHNIQUE

This technique, the same as the one based on estimation of the Walsh spectrum coefficients, is an expansion of syndrome testing and is based on the analysis of the autocorrelation function $B(T)$ in a Boolean expression $F(x_1,\ldots,x_n)$ that describes the behaviour of a circuit under test. Thus, by function $B(T)$, where T is the vector of binary variables $\tau_1, \tau_2, \ldots, \tau_n, \tau_i \in \{0, 1\}$, $i = 1$ to n, we understand the function [140].

$$B(\tau_1,\ldots,\tau_n) = \sum_{x_1\cdots x_n=0\cdots0}^{1\cdots1} F(x_1,\ldots,x_n)F(x_1\oplus\tau_1,\ldots,x_n\oplus\tau_n) \qquad (10.11)$$

which takes on arbitrary values from the set $\{0, 1, 2, \ldots, 2^n\}$ depending on specific values of $\tau_1, \tau_2, \ldots, \tau_n$. As an example of calculating $B(T)$ by (10.11), let us consider the procedure of autocorrelation function evaluation for $F(x_1, x_2, x_3) = x_2(x_1 + \bar{x}_3)$ with $\tau_1\tau_2\tau_3 = 100$. Then we obtain

$$B(100) = \sum_{x_1x_2x_3=000}^{111} F(x_1, x_2, x_3)F(x_1\oplus1, x_2\oplus0, x_3\oplus0)$$

$$= \sum_{x_1x_2x_3=000}^{111} F(x_1, x_2, x_3)F(\bar{x}_1, x_2, x_3)$$

$$= \sum_{x_1x_2x_3=000}^{111} x_2(x_1, \bar{x}_3)x_2(\bar{x}_1 + \bar{x}_3)$$

$$= \sum_{x_1x_2x_3=000}^{111} x_2\bar{x}_3 = 2.$$

The technique that implements the analysis of autocorrelation function $B(T)$ is based on the use of the following theorem [141].

Theorem 10.7. A two-level non-redundant digital circuit defined by a Boolean function $F = G_1x_i + G_2\bar{x}_i + G_3, i\in\{1, 2, 3, \ldots, n\}$, where $G_1 \neq 0$, $G_2 \neq 0$, $G_3 \neq 1$, $\partial G_1/\partial x_i = \partial G_2/\partial x_i = \partial G_3/\partial x_i = 0$, is correlation-testable with respect to the input applied to x_i on the basis of analysis $B(\tau_i,\ldots,\tau_n)$, where $\tau_i = 1$, $\tau_j = 0$ for all $j \neq i\in\{1, 2, 3, \ldots, n\}$.

Proof. By substituting $F(x_1,\ldots,x_n) = G_1x_i + G_2\bar{x}_i + G_3$ and $F(x_1\oplus\tau_1,\ldots,x_n\oplus\tau_n) = F(x_1,\ldots,\bar{x}_i,\ldots,x_n) = G_1\bar{x}_i + G_2x_i + G_3$ into (10.11), we obtain

$$B(T_i) = \sum_{x_1 \cdots x_n = 0 \cdots 0}^{1 \cdots 1} F(x_1, \ldots, x_i, \ldots, x_n) F(x_1, \ldots, \bar{x}_i, \ldots, x_n)$$

$$= \sum_{x_1 \cdots x_n = 0 \cdots 0}^{1 \cdots 1} (G_1 G_2 + G_3)$$

where T_i is the vector $\tau_1 \cdots \tau_i \cdots \tau_n = 0 \cdots 1 \cdots 0$ for which the condition $\tau_i = 1$ is satisfied and $\tau_j = 0$ for all $j \neq i \in \{1, 2, 3, \ldots, n\}$. From the expression obtained, it follows that $B(T_i) = S(G_1 G_2 + G_3)$, where $S(G_1 G_2 + G_3)$ is the syndrome of function $G_1 G_2 + G_3$ with no account of the constant factor $1/2^n$ in the expression (4.7). Then the syndrome $S(F)$ for the desired function $F = G_1 x_i + G_2 \bar{x}_i + G_3$ may be calculated as

$$S(F) = S(G_1 x_i + G_2 \bar{x}_i + G_3). \tag{10.12}$$

When the fault $x_i \equiv 0$ occurs, the autocorrelation function $B(T_i^0)$ is determined by

$$B(T_i^0) = \sum_{x_1 \cdots x_n = 0 \cdots 0}^{1 \cdots 1} (G_2 + G_3)(G_2 + G_3)$$

$$= \sum_{x_1 \cdots x_n = 0 \cdots 0}^{1 \cdots 1} (G_2 + G_3) = S(G_2 + G_3)$$

where $S(G_2 + G_3)$ is, by (10.2), a syndrome of F_0^*, which describes the behaviour of the circuit implementing F for $x_i \equiv 0$. Thus $B(T_i^0) = S(F_0^*)$. For $x_i \equiv 1$, we may similarly show that

$$B(T_i^1) = = \sum_{x_1 \cdots x_n = 0 \cdots 0}^{1 \cdots 1} (G_1 + G_3)(G_1 + G_3) = S(G_1 + G_3) = S(F_1^*).$$

Let us compare the obtained values of correlation functions $B(T_i^0)$ and $B(T_i^1)$ for faulty modifications of the circuit that corresponds to $F(x_1, \ldots, x_n)$ with $B(T_i) = S(G_1 G_2 + G_3) = S[(G_1 + G_3)(G_2 + G_3)]$, which characterizes its fault-free state. For this purpose, we shall use the syndrome product property of the two functions described by the relation (4.10)

$$S(F_1 F_2) = S(F_1) - S(F_1 \bar{F}_2).$$

From the obtained relation for $F_1 \neq F_2$ and $F_1 \neq \bar{F}_2$, it follows that

$$S(F_1 F_2) < S(F_1), S(F_2). \tag{10.13}$$

The Boolean functions $(G_1 + G_3)$ and $(G_2 + G_3)$ satisfy the conditions of (10.13). In fact, $G_1 + G_3 \neq G_2 + G_3$, since $G_1 \neq G_2$ and $G_1 + G_3 \neq \overline{G_2 + G_3}$, since $(G_1 + G_3) + (G_2 + G_3) \neq 0$ [141]. Thus, by using (10.13), we may demonstrate that

$$B(T_i^0) = S(G_2 + G_3) > B(T_i) = S[(G_1 + G_3)(G_2 + G_3)]$$

and

$$B(T_i^1) = S(G_1 + G_3) > B(T_i) = S[(G_1 + G_3)(G_2 + G_3)].$$

Hence the fault that has occurred at the input x_i to the circuit that implements the Boolean function $F(x_1, \ldots, x_n)$ will be detected by testing the autocorrelation function $B(T_i)$ and thus it is correlation-testable, as was to be shown. ∎

The basic results of Theorem 10.7 may be used to provide for testability of syndrome-untestable faults. Figure 4.3 shows an example of fault $x_1 \equiv 1$ in a circuit that implements the switching function $F = x_1 \bar{x}_2 \bar{x}_3 + \bar{x}_1 x_2 x_3$. For the specified fault, the syndrome $S(F) = S(x_1 \bar{x}_2 \bar{x}_3 + \bar{x}_1 x_2 x_3)$ equals $S(F_1^*) = S(\bar{x}_2 \bar{x}_3)$, thereby implying that the fault $x_1 \equiv 1$ is syndrome-untestable. At the same time, the value

$$B(T_1) = \sum_{x_1 x_2 x_3 = 0 \cdots 0}^{111} (x_1 \bar{x}_2 \bar{x}_3 + \bar{x}_1 x_2 x_3)(\bar{x}_1 \bar{x}_2 \bar{x}_3 + x_1 x_2 x_3) = 0$$

differs from

$$B(T_1^1) = \sum_{x_1 x_2 x_3 = 000}^{111} \bar{x}_2 \bar{x}_3 = 2$$

for function $F = x_1 \bar{x}_2 \bar{x}_3 + \bar{x}_1 x_2 x_3$, thereby implying that the specified fault is testable by autocorrelation function analysis (10.11).

In [141] is presented an example of correlation technique application for the syndrome-untestable node of the digital circuit under test. Unlike the techniques that have been discussed in Sec. 4.4, correlation-testability of the circuit requires no extra logic and external nodes. However, in order to implement the technique discussed, it is necessary to use more complicated hardware for carrying out the test equipment, which is more time-consuming than syndrome testing. Actually, the block diagram of correlation testing consists of a number of blocks specific for the technique. Thus apart from the test pattern generator, circuit under test and binary counter, which are conventional units of the tester, the device configuration also includes an assembly of M2 gates, $\tau_1 \cdots \tau_n$ code generator, demultiplexer, delay unit, two-input AND gate, second counter and control unit [141]. The $\tau_1 \cdots \tau_n$ code generator successively generates the code $\tau_1 \cdots \tau_n$ with $\tau_j = 0$, $j = 1$ to n, and $\tau_i = 1$ and $\tau_j = 0$ for all $j \neq i \in \{1, 2, 3, \ldots, n\}$. In so doing, the inputs of the circuit under test are successively applied with the patterns of variables $x_1 \cdots x_i \cdots x_n$ and $x_1 \cdots x_i \cdots x_n$. The output value $F(x_1, \ldots, x_i, \ldots, x_n)$ of the circuit is applied to the counter input and delayed by a cycle at the delay element. Thus, the product $F(x_1, \ldots, x_i, \ldots, x_n) F(x_1, \ldots, \bar{x}_i, \ldots, x_n)$ is produced at the output of the two-input AND gate. As a result, the first counter will contain the value of syndrome $S(F)$ for $F(x_1, \ldots, x_n)$ and the second counter will contain the value of autocorrelation function $B(T_i)$.

Theorem 10.7 reveals the possibilities of correlation testing for the specific case, i.e. code $\tau_1 \cdots \tau_n$, when only one of its coordinates is 1, and demonstrates that the circuit can be made testable without modification. However, the desired testability will be attained by introducing extra values of $B(T_i)$. Their number may reach n together with syndrome $S(F)$. Therefore, it is not reasonable to implement both correlation and spectral techniques without special strategies of digital circuit design, which take account of the specific character of test response generation. However, the urgency of further investigations in both techniques discussed with a view to solving the diagnostic test problems should be noted.

10.4. COMPRESSION RESULT AS A SUM OF CURRENT SYNDROME VALUES

One way to extend syndrome testing is to use both the syndrome (Walsh spectrum coefficient r_0) and the characteristic, which represents the process of its generation as the result of testing [142]. In this case, the syndrome is taken to mean the value $S(y)$ defined by (4.2) and indicating the number of 1s in the sequence $\{y(k)\}$. Then the current value of syndrome $S_j(y)$ in the jth cycle of its generation will be determined by the expression

$$S_j(y) = \sum_{k=1}^{j} y(k) \qquad j = 1 \text{ to } l \tag{10.14}$$

where $y(k) \in \{0, 1\}$ is the kth symbol of the sequence $\{y(k)\} = y(1)y(2) \cdots y(l)$; $S(y)$ is the final value of $S(y)$ whose generation process is determined by (10.14).

In [142] it has been proposed that the value of A,

$$A = \sum_{j=1}^{l} S_j(y) = \sum_{j=1}^{l} \sum_{k=1}^{j} y(k) = \sum_{j=1}^{l} (l + 1 - j)y(j) \tag{10.15}$$

be used as an integral characteristic that allows one to estimate the progress of $S(y)$ generation.

The use of syndrome $S(y)$ and integral characteristic A gave rise to the new compact testing method based on accumulation of compression results (accumulator compression testing) [142]. The block diagram of the device consisting of a test sequence generator, a circuit under test, a counter and an adder reveals the essence of this testing technique. The contents of the counter and adder, which represent syndrome $S(y)$ and integral characteristic A, will be used as a signature. Consider the case of a sequence $\{y(k)\}$ that contain $l/2$ 1s, which is the worst one for syndrome testing since the value $C_\alpha^l - 1$ indicates the number of sequences $\{y(k)\}$ that contain error symbols, have signature identical to the expected one and reach their maximum with $\alpha = l/2$. For these sequences, the progress of signature generation differs from that of expected signature generation; however, the final value equals $S(y)$, i.e. the expected signature. An integral characteristic that allows one to estimate the signature generation process is the value A (see equation (10.15)) representing the sum of numbers each of which is less than or equal to l. The number of sequences $\{y(k)\}$ that have identical A and $S(y)$ equals the number of partitions resulting from division of A by h such that each of h partitions is less than or equal to l. Represent the number of such sequences as a function $D[S(y), l, A]$ whose arguments are related by

$$S(y)[S(y) + 1]/2 \leqslant A \leqslant l(l + 1)/2 - [l - S(y)][l - S(y) + 1]/2 \tag{10.16}$$

where $S(y)[S(y) + 1]/2$ is defined by the sequence $\{y(k)\}$ of the form

$$00 \cdots 0\underbrace{01 \cdots 1}_{S(y)} \tag{10.17}$$

and equals the minimal value of A for sequences with syndrome $S(y)$. The expression $l(l + 1)/2 - [l - S(y)][l - S(y) + 1]/2$ is the maximal value of A for the sequence $\{y(k)\}$ which comprises $S(y)$ 1s. This sequence has the following form

$$\underbrace{11 \cdots 1}_{S(y)}000 \cdots 0.$$

Subject to (10.16), the following theorem can be proved for the function $D[S(y), l, A]$ [142].

Theorem 10.8. The number of sequences $\{y(k)\}$ of length l, which have identical values of syndrome $S(y)$ and integral characteristic A, is determined by

$$D[S(y), l, A] = N\{S(y), l - S(y), A - S(y)[S(y) + 1]/2\}$$

where $N\{S(y), l - S(y), A - S(y)[S(y) + 1]/2\}$ is the number of partitions $A - S(y)$ $[S(y) + 1]/2$ by at most $S(y)$ partitions each of which is less than or equal to l.

Proof. Any sequence $\{y(k)\}$ of length l that contains $S(y)$ 1s and differs from the expected sequence will be untestable by syndrome $S(y)$ and integral characteristic A if it has the same value of A or A_1 defined by the expression

$$A_1 = A - S(y)[S(y) + 1]/2.$$

The value of A_1 various from 0 to $l(l + 1)/2 - [l - S(y)][l - S(y) + 1]/2 - S(y)[S(y) + 1]/2$ with the zero value of A_1 having the sequence $\{y(k)\}$, which corresponds to (10.17). For the sequences obtained from (10.17), identical values of A_1 are achieved by repositioning its unit symbols.

Any new position of a 1 will be associated with the number in the expression (10.15) defining A, which does not exceed $S(y)$. Furthermore, the total number of position changes for the 1s must not exceed $S(y)$ when a sequence $\{y(k)\}$ is being formed from the basic sequence obtained by (10.17). Thus, we may conclude that $D[S(y), l, A]$ is the number of position changes for 1s, which does not exceed $S(y) (A - S(y)[S(y) + 1]/2$ partitioning by at most $S(y))$, with each position being associated with a number that is less than or equal to $l - S(y)$,

$$D[S(y), l, A] = N\{S(y), l - S(y), A - S(y)[S(y) + 1]/2\}$$

as was to be shown. ∎

As has been shown in [142], the maximum value of function $D[S(y), l, A]$ may be reached only for $S(y) = 1/2$ and $A_1 = \text{int}[S^2(y)/2]$. Similarly, the maximum number of undetectable error sequences will be determined by function $N[S(y), S(y), S^2(y)/2]$, which is the upper bound on the number of undetectable violations in the sequence $\{y(k)\}$.

For each specific value $S(y)$, the function $N[S(y), S(y), S^2(y)/2]$ assumes the value that may be estimated by the asymptotic formula

$$N[S(y), S(y), S^2(y)/2] \simeq \frac{\sqrt{3} \times 4^{S(y)}}{\pi S^2(y)} \tag{10.18}$$

obtained in [142].

A correspondence between the formula (10.18) and the precise value of function $N[S(y), S(y)/2]$ is proved by practical results (Table 10.2 [142].

The number of error sequences $\{y(k)\}$ that are undetectable by $S(y)$ and A and computed by (10.18) may be estimated by

$$2^{l - 2\log_2 l} \tag{10.19}$$

where $l = 2S(y)$; the value of $\sqrt{3}/\pi$ is set equal to 1 [142]. The analysis of the latter expression shows that the number of errors that cannot be detected by the technique discussed exceeds the number of errors undetectable by a signature analyser

Table 10.2.

$S(y)$	$N[S(y), S(y), S^2(y)/2]$	$\sqrt{3} \times 4^{S(y)}/\pi S^2(y)$
13	208960	218929.32
28	4.95×10^{13}	5.06×10^{13}
46	1.27×10^{24}	1.29×10^{24}
53	1.57×10^{28}	1.59×10^{28}
62	7.80×10^{32}	7.87×10^{32}

Table 10.3.

Circuit under test	Faults detectable by $S(y)$ and A (%)	Faults detectable by $S(y)$ (%)
Arithmetic and logical unit	100	94.1
	100	99.9
	100	100
Four-bit adder	99.1	85.5
	100	81.1
	100	89.4
	100	84.5
Decoding circuit	100	100
	100	100

implemented by a shift register of $2\log_2 l$ in length. It should be noted, however, that the value obtained by (10.19) is only one less than the respective value inherent to signature analysis.

Table 10.3 summarizes comparative results for the discussed technique and signature analysis. From the table, it follows that the technique based on the analysis of syndrome $S(y)$ and integral characteristic A has the advantage over signature analysis.

10.5. COMPRESSION IN SPACE AND TIME

All the compact testing methods that have just been discussed are directed towards solving the problem of compressing long output responses of a digital circuit into short keywords. The possibility of such a procedure is based on the rather realistic hypothesis that the final number of symbols in the output sequence of the circuit under test might be distorted. The hypothesis allows one to estimate the occurrence of a fault in the circuit by an integral characteristic of its output response. In this case, the length l of sequence being analysed normally far exceeds the length of keyword m; however, the test coverage obtained will be high.

The procedure of compressing output responses of digital circuits is by its very nature the procedure of compression in time. An example of the procedure is syndrome testing implementation, which consists of adding the current symbol of the sequence under test to the sum of symbols obtained at previous instants of time. When used in practice, compression in time allows one to gain significant advantages from reduction in test data volume, application of long test sequences, etc. Actually, every compact testing method is oriented to a single-output digital circuit. Therefore, a multi-output circuit is normally tested at each output by a unified scheme of compression method implementation. However, for a v-output digital circuit, we may suppose that the circuit faults will be observable by distortion of $w \leqslant v$ output responses. This is supported by the example of a v-output digital circuit whose outputs are associated with the specific set of its gates and inputs. The specified sets and the asssociated sets for other circuit outputs are disjoint. In this case any single error in the circuit will cause only one output sequence to be distorted. In the general case, the number of output responses with error characters will depend on the number of circuit faults, the type of test sequence used and the level of functional dependence between the sequence produced at its outputs. However, we may always assume that any fault(s) in the circuit with v outputs will always be observable at least at $w \leqslant v$ outputs of the circuit. This makes it possible for the character values to be tested by an integral characteristic at v circuit outputs at a time. An example of such a test is the case of a v-output digital circuit whose faults manifest themselves at $w = 1$ outputs only. Then their appearance will be observable by testing only one sequence produced as a modulo-2 sum of v output responses of the circuit. In practice, the implementation of the circuit that generates the summed sequence will consist of the use of a v-input modulo-2 adder.

The example just discussed has demonstrated the procedure of compression in space [143], which consist of reducing the amount of output responses to be tested in the circuit, which are then compressed in time to obtain the final characteristic in the form of a keyword. In the general case, compact testing that realizes two-stage compression is performed by a special device. Such a device may be a circuit for compression in space, which allows one to reduce the number of tested sequences to $u \leqslant v$, to be further compressed in time to obtain an m-bit keyword. Then the value u, which defines the number of outputs in a circuit producing the integral characteristic, is a function of v and w, i.e. $u = f(v, w)$. If the value v is uniquely characterized by the number of circuit outputs, it will be rather difficult to estimate w, either by simulating all possible circuit faults or by analysing the circuit structure. We may practically always estimate the maximum value of w, which is known to be greater than its actual value but less than the number of circuit outputs v. Thus the problem arises of designing a device that converts v output sequences of the digital circuit with a fault manifesting itself on w outputs at least into $u = f(v, w)$ sequences containing data on the fault that has occurred. Then the primary requirement imposed on the device is its ease of implementation, which resides in the fact that the number of elements to be used for obtaining the minimal value of u should be small. Any of the discussed compact testing methods whose hardware implementation allows compression in space may be a candidate for designing such a device. However, the scheme of linear conversion of v input sequences into u output sequences, which is based on coding theory concepts, is more suitable [144]. In this case, the scheme realizing the procedure of compression in space will perform conversion associated with testing matrix H of linear code $(v, v - u, w + 1)$. The basic characteristics of such a scheme are determined by the following theorem [143].

Table 10.4.

v	w	$u = f(v, w)$ min	max	Code applied
v	1	1	1	Parity checking
$2^m - 1$	2	m	m	Hamming code $(2^m - 1, 2^m - 1 - m, 3)$
23	6	11	11	Golay code (23, 12, 7)

Theorem 10.9. The data compression scheme that realizes the testing of matrix H of linear code $(v, v - u, w + 1)$ with v inputs and u outputs allows one to detect all errors whose multiplicity does not exceed w.

The proof of the above theorem follows uniquely from the property of the linear code $(v, v - u, w + 1)$, which includes $(v - u)$ data characters, total code word length of v and code distance $w + 1$, to detect all possible errors of multiplicity less than $w + 1$.

An example of a linear code that allows one to detect any single error is the code (7, 6, 2) described by the testing matrix

$$H = |1 \quad 1 \quad 1 \quad 1 \quad 1 \quad 1 \quad 1|.$$

Implementation of the compression scheme based on matrix H has seven inputs and only one output. A more complex example is the use of code (7, 3, 3) whose testing matrix appears as

$$H = \begin{vmatrix} 0 & 0 & 0 & 1 & 1 & 1 & 1 \\ 0 & 1 & 1 & 0 & 0 & 1 & 1 \\ 1 & 0 & 1 & 0 & 1 & 0 & 1 \end{vmatrix}. \tag{10.20}$$

The scheme of a device that performs compression in space and is based on matrix (10.20) has seven inputs and three outputs. It provides for detecting all errors whose multiplicity does not exceed 2.

In [143] are presented the maximum and minimum estimates for $u = f(v, w)$, which are summarized in Table 10.4 for the most commonly used codes. As seen from the table, the Golay code allows one to construct compression schemes with 23 inputs and 11 outputs. In so doing, the volume of test data is decreased by $23 - 11 = 12$ output responses of a digital circuit of length l. At the same time, provision is made for detecting all faults that manifest themselves at no more than six outputs from the circuit.

The proposed procedure of compressing data in space can be built into test equipment or LSI chip [143]. By this means the amount of test data to be analysed in the course of a test experiment may be significantly reduced. However, the efficiency of such an approach is not high without the procedure of compression in time. Therefore, the most commonly used practical solutions are based on the use of multi-line time compression schemes, which perform data compression in space and time concurrently. We next consider the basic methods of constructing such schemes.

11

ANALYSIS OF MULTI-OUTPUT DIGITAL CIRCUITS

11.1. DESIGN OF MULTI-LINE ANALYSING CIRCUITS

The problem of analysing multi-output digital circuits in the course of their testing consists of detecting a fault by the output responses of the circuit. A characteristic property of such analysis is the need to test a large amount of output responses of the circuit (it may run into hundreds). Therefore, in this case, conventional compact testing methods used for single-output digital circuits would not do the job. In fact, any attempt to analyse an n-output digital circuit by a series signature analyser either increases the circuit testing time by n or requires hardware for implementing n signature analysers. The question of signature length, which may be made n times as large, is left open. However, the use of single-line signature analysers allows one to find a compromise solution which is based on the use of only $u < n$ analysers. Despite this fact, however, special methods and techniques are more often used in practice. Among them the most commonly used is the method based on transformation of n output sequences $y_i(k) \in \{0, 1\}$, $i = 1$ to n, $k = 1$ to l, of length l into a single sequence $\{y_0(k)\}$ by the expression

$$y_0(k) = \sum_{i=1}^{n} y_i(k) \qquad k = 1 \text{ to } l. \tag{11.1}$$

The method can be implemented as a procedure of compression in space and time. In either of the cases the idea of obtaining compact estimates intrinsic to compact testing methods is implemented.

To measure the efficiency of a data compression algorithm that implements the relation (11.1), the compact testing methods (see Sec. 8.5) have been estimated by comparison. The principle of the method is the comparative analysis of probability distribution P_n^μ of failing to detect an error of multiplicity $\mu \in \{1, 2, 3, ..., l\}$, where l is the length of sequence under test. In this case, P_n^μ is evaluated as the ratio of undetectable μ-multiple errors to the total number of possible errors containing μ error symbols.

We find the values of probability P_n^μ of failing to detect an error of multiplicity $\mu \in \{1, 2, 3, ..., nl\}$ that has occurred in n output sequences of length l as a result of their transformation by (11.1). For $\mu = 1$, any single error that affected a symbol in one of the n output sequences $y_i(k) \in \{0, 1\}$, $i = 1$ to n, $k = 1$ to l, will invert one symbol in the result of compressing these sequences. Thus, we may deduce that any error of multiplicity $\mu = 1$ will be detectable by the compression method discussed. At the same time, for $\mu = 2$ it might occur that double errors will remain undetectable as follows from relation (11.1). In fact, when the actual values $y_j^*(k)$ and $y_d^*(k)$, $j \neq d \in \{1, 2, 3, ..., n\}$, differ from the expected values $y_j(k)$ and $y_d(k)$, the actual value of symbol $y_0^*(k)$ in the resulting sequence $\{y_0(k)\}$ will match the expected value $y_0(k)$. The number of such situations is determined by the relation

$$V_n^2 = C_l^1 C_n^2$$

and the total number V^2 of all possible errors of multiplicity $\mu = 2$ is C_{ln}^2. Thus, the probability P_n^2 is calculated as

$$P_n^2 = \frac{C_l^1 C_n^2}{C_n^2} = \frac{n-1}{ln-1}. \tag{11.2}$$

Considering that the length l of sequence under test normally satisfies the condition $l \gg 1$, the expression (11.2) may be rearranged as

$$P_n^2 = \frac{n-1}{n}\frac{1}{l}. \tag{11.3}$$

For $\mu = 3$ as well as for any odd value of μ, $P_n^\mu = 0$, since any violation in the odd number of symbols in sequences $\{y_i(k)\}$ will always cause a difference between the actual sequence $\{y_c^*(k)\}$ and its expected form $\{y_0(k)\}$ as follows from (11.1). The same relation, however, shows that for any even μ, $P_n^\mu \neq 0$. Thus, for $\mu = 4$, P_n^4 is calculated by

$$P_n^4 = \begin{cases} \dfrac{C_n^2 C_n^2 C_l^2}{C_{nl}^4} & \text{for } n < 4 \\[3mm] \dfrac{C_n^2 C_n^2 C_l^2}{C_{nl}^4} + \dfrac{C_l^1 C_n^4}{C_{nl}^4} & \text{for } n \geq 4 \end{cases} \tag{11.4}$$

which may be transformed for $n \geq 4$ into

$$\frac{3n(n-1)^2(l-1)}{(nl-1)(nl-2)(nl-3)} + \frac{(n-1)(n-2)(n-3)}{(nl-1)(nl-2)(nl-3)}$$

where $l \gg 1$. Substituting $(nl-1)$ and $(l-1)$ by nl and l, we finally obtain

$$P_n^4 = \begin{cases} 3\left(\dfrac{n-1}{n}\right)^2 \dfrac{1}{l^2} & \text{for } n < 4 \\[3mm] 3\left(\dfrac{n-1}{n}\right)^2 \dfrac{1}{l^2} + \dfrac{(n-1)(n-2)(n-3)}{n^3}\dfrac{1}{l^3} & \text{for } n \geq 4. \end{cases} \tag{11.5}$$

Ignoring the second term for the case of $n \geq 4$, we may evaluate P_n^4 by

$$P_n^4 = 3\left(\frac{n-1}{n}\right)^2 \frac{1}{l^2}. \tag{11.5}$$

Similarly to relation (11.4) for $\mu = 4$, we may find P_n^μ for $\mu = 6$ as

$$P_n^6 = \begin{cases} \dfrac{C_n^2 C_n^2 C_n^2 C_l^2}{C_{nl}^6} & \text{for } n < 4 \\[3ex] \dfrac{C_n^2 C_n^2 C_n^2 C_l^3 + C_n^2 C_n^4 C_l^2}{C_{nl}^6} & \text{for } 4 \leqslant n < 6 \\[3ex] \dfrac{C_n^2 C_n^2 C_n^2 C_l^3 + C_n^2 C_n^4 C_l^2 + C_n^6 C_l^1}{C_{nl}^6} & \text{for } n \geqslant 6. \end{cases}$$

Under the assumptions that have been made when deriving equation (11.5), we may rearrange it as

$$P_n^6 = 15 \left(\frac{n-1}{n} \right)^3 \frac{1}{l^3}. \tag{11.6}$$

Generalizing equations (11.3), (11.5) and (11.6), we may obtain the resulting expression to calculate the probability distribution P_n^μ for even μ:

$$P_n^\mu \simeq \prod_{j=1}^{\mu/2} (2j-1) \left(\frac{n-1}{n} \right)^{\mu/2} \frac{1}{l^{\mu/2}}. \tag{11.7}$$

Note also that for odd μ, $P_n^\mu = 0$.

To estimate the form of fault escape probability distribution P_n^μ by the method based on transforming n sequences $\{y_i(k)\}$ into a single resulting sequence $\{y_0(k)\}$ by (11.1), let us consider an example for a three-output ($n = 3$) circuit whose output responses are $l = 21$ long. Owing to transformation of the three initial sequences $\{y_1(k)\}$, $\{y_2(k)\}$ and $\{y_3(k)\}$ into a single sequence $\{y_0(k)\}$, $k = 1$ to 21, some of the errors became undetectable. This will be evaluated in the integral form by

$$P_n^\mu = 0 \qquad \mu = 1, 3, 5, \ldots$$

$$P_n^\mu = \prod_{j=1}^{\mu-1} (2j-1) \left(\frac{2}{3} \right)^{\mu/2} \frac{1}{21^{\mu/2}} \qquad \mu = 2, 4, 6, \ldots \tag{11.8}$$

which is true for values of μ less than the sequence length l. Putting a bound on $\mu = 6$, define the proobability P_n^μ by (11.8). Then we obtain that $P_n^1 = P_n^3 = P_n^5 = 0$ and $P_n^2 = 0.0317$, $P_n^4 = 0.0030$ and $P_n^6 = 0.0005$. For $l = 5$ the associated probability values will take the form $P_n^1 = P_n^3 = P_n^5 = 0$ and $P_n^2 = 0.142$, $P_n^4 = 0.066$ and $P_n^6 = 0.054$, although more precise expressions for P_n^μ rather than the approximate relation (11.7), which holds true only for $\mu \ll l$, have been used here.

The analysis of the obtained P_n^μ evaluations as well as of their general expression (11.7) shows the non-uniformity of the distribution, which allows one to comment on the rather moderate efficiency of the discussed compression algorithm (see Sec. 9.1). It should be noted, moreover, that the compression result is as large as the length l of circuit output response, which is prohibitive. Therefore a trade-off is most often used in practice, which consists of two-step conversion of output responses from an n-output digital circuit. First of all, n output sequences $\{y_i(k)\}$ of length l are converted into a sequence $\{y_0(k)\}$ by (11.1). Then the probability P_{d1}^μ of detecting an error of multiplicity $\mu \in \{1, 2, 3, \ldots, nl\}$ that has occurred in the initial sequences $\{y_i(k)\}$ is determined by

$$P^\mu_{d1} = 1 - P^\mu_{n1} \tag{11.9}$$

where the values of P^μ_{n1} are calculated by the discussed technique above, and for $\mu \leqslant l$ which assume an even value they are defined by (11.7). Then the resulting sequence $\{y_0(k)\}$ is compressed into an m-bit keyword by one of the compact testing techniques. For the case of signature analysis, the two-step conversion scheme is shown in Fig. 11.1. To estimate the validity of conversion, let us estimate the probability distribution P^μ_{n0} of failing to detect an error in sequences $\{y_i(k)\}$ by testing their signature $S(x)$ produced by the shift-register memory elements (Fig. 11.1). Let $L < 2^m$ and the events of error detection in sequences $\{y_i(k)\}$ by analysing $\{y_0(k)\}$ and in the sequence $\{y_0(k)\}$ by analysing the m-bit signature $S(x)$ be independent. Then assuming that $P^\mu_{n2} = 1/2^m$ for any μ, we obtain

$$P^\mu_{d0} = 1 - P^\mu_{n0} = (1 - P^\mu_{n1}) - (1 - 1/2^m).$$

Hence

$$P^\mu_{n0} = P^\mu_{n1} + \frac{1}{2^m} - P^\mu_{n1}\frac{1}{2^m}. \tag{11.10}$$

The expression for P^μ_{n0} is rather approximate in nature; nevertheless it allows one to evaluate the efficiency of the two-step compression scheme, which is an example of implementing the concept of compression in time and space (see Sec. 10.5). The first and the most evident result from analysing the obtained expression for P^μ_{n0} may be a conclusion on non-uniformity of the P^μ_{n0} distribution. In fact, for $\mu = 3$, $P^3_{n0} = 1 \times 2^m$, since $P^3_{n1} = 0$; and for $\mu = 4$, P^4_{n0} will be define by

$$P^4_{n0} = 3\left(\frac{n-1}{n}\right)^2\frac{1}{l^2} + \frac{1}{2^m} - 3\left(\frac{n-1}{n}\right)^2\frac{1}{l^2}\frac{1}{2^m}$$

where the final value of P^4_{n0} depends on the ratio of n, l and $2m$. For the most realistic case when $nl \simeq 2^m$

$$P^4_{n0} = \frac{3(n-1)^2}{2^{2m}} + \frac{1}{2^m} - \frac{3(n-1)^2}{2^{3m}}.$$

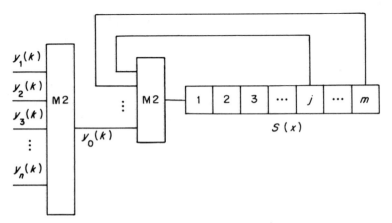

Fig. 11.1. Block diagram of two-step conversion of compressed sequences

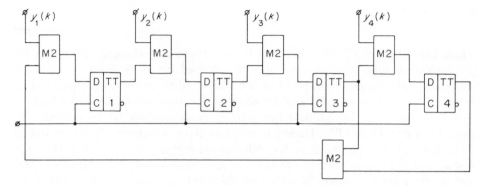

Fig. 11.2. A four-input signature analyser

Hence for $m = 16$ and $n = 196$ we shall have, for example, $P_{n0}^4 \simeq 3/2^m$ which is three times as large as P_{n0}^3.

The most widespread structure of a parallel signature analyser applied to testing multi-output digital circuits is the BILBO (see Sec. 3.5), which is often used for implementing a VLSI self-test. An example of a four-input analyser based on polynomial $\varphi(x) = 1 \oplus x^3 \oplus x^4$ is shown in Fig. 11.2. It is used for the analysis of output responses of four-output digital circuits. For the sequences $\{y_i(k)\}$, $i = 1$ to 4, whose forms are summarized in Table 11.1, the codes produced by the analyser's memory elements are given (Fig. 11.2).

Then the final value of code $a_1(k)a_2(k)a_3(k)a_4(k)$ will be the resulting value of signature $S(y)$, which is the compact estimate of compressing the four sequences $\{y_i(k)\}$. However, the same result might be obtained by using the structure of two blocks that explicitly correspond to the algorithms of compression in space and time. In fact, considering that $y_4(k) = 0$ for $k = 0$ to 5, we may show the equivalence of schemes shown in Figs 11.2 and 11.3. Thus for the scheme of Fig. 11.3, the input sequence $\{y_0(k)\}$ of a serial signature analyser, which is produced on the output of the first modulo-2 adder, is defined by

$$y_0(k) = \sum_{i=1}^{3} y_i(k - 3 + i) \qquad k = 0 \text{ to } 5 \qquad (11.11)$$

Table 11.1.

h	$y_1(k)$	$y_2(k)$	$y_3(k)$	$y_4(k)$	$a_1(k)$	$a_2(k)$	$a_3(k)$	$a_4(k)$
0	0	0	0	0	0	0	0	0
1	1	0	1	0	1	0	1	0
2	1	1	1	0	0	0	1	1
3	0	0	0	0	0	0	0	1
4	1	0	0	0	0	0	0	0
5	1	1	1	0	1	1	1	0

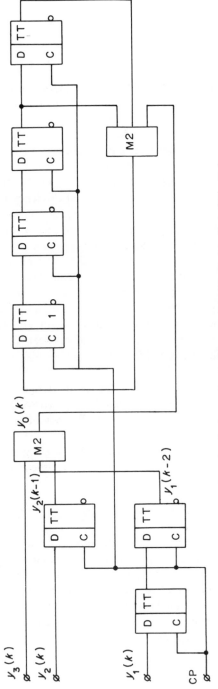

Fig. 11.3. Functional scheme of a signature analyser identical to that of Fig. 11.2

where $y_1(k-1) = y_2(k-1) = y_1(k-2) = 0$. Using the relation (11.11), we obtain

$$
\begin{array}{lccccc}
y_1(k) = & & 1 & 1 & 0 & 1 & 1 \\
y_2(k) = & 0 & 1 & 0 & 0 & 1 \\
y_3(k) = 1 & 1 & 0 & 0 & 1 \\
\hline
y_0(k) = 1 & 1 & 0 & 1 & 1 & 0 & 1
\end{array}
$$

The result of compressing the obtained sequence $\{y_0(k)\}$, $k = 1$ to 7, will be identical to the value of signature $S(y)$, thereby proving the equivalence of the discussed schemes as to the final result.

The given example shows that the widespread scheme of parallel signature analyser (BILBO) is practically equivalent to a simple scheme of two-step data compression (Fig. 11.1).

Although generation of the resulting sequence $\{y_0(k)\}$ by the BILBO appproach is more complex than by (11.1), the earlier discussed efficiency measurement technique is used in both cases. According to this technique, non-uniformity of probability distribution P_n^μ of failing to detect an error of multiplicity μ is characteristic of both approaches. Besides, the structure of a parallel signature analyser (Fig. 11.2), like the length of its signature $S(y)$, is uniquely defined by the number of outputs n in the circuit under test. Therefore, with $n > 100$, the complexity of compressor and the number of bits in the signature $S(y)$ became prohibitive. An attempt to cascade the parallel signature analysers (PSAs) allows one to minimize the resulting structure size; however, the efficiency of such an analyser is difficult to estimate.[145, 146], since it will depend on the layout of connections between PSAs and their particular implementation. The efficiency of PSA based on an n-bit adder is also difficult to measure.

Thus it has been found more advantageous to design PSAs by using the well studied techniques for designing serial signature analysers (SSA). Owing to this, the SSA theory might be used for evaluating the efficiency of parallel signature analysis. An example of such a solution is the PSA design technique based on primitive polynomials $\varphi(x)$ [147], which involves the use of coding theory for evaluating the efficiency of the PSA-based analysis. Let us consider the concepts of this technique.

11.2. DESIGN OF PARALLEL SIGNATURE ANALYSERS

To design parallel signature analysers it is preferable to use the technique of designing PSAs with a random number of inputs and an independent set of memory elements defined by the high degree of the generating polynomial $\varphi(x)$ only [147]. This technique is based on the primitive polynomial $\varphi(x) = 1 \oplus \alpha_1 x \oplus \alpha_2 x^2 \oplus \cdots \oplus \alpha_m x^m$, where $m = \deg \varphi(x)$ defines the analysis validity, and the length of signatures generated.

As has been shown earlier, for a $\varphi(x)$ the behaviour of a serial signature analyser is described by the set of equations

$$a_i(0) = 0 \qquad i = 1 \text{ to } m$$

$$a_1(k) = y(k) \oplus \sum_{i=1}^{m} \alpha_i a_i(k-1) \tag{11.12}$$

$$a_j(k) = a_{j-1}(k-1) \qquad j = 2 \text{ to } m, \, k = 1 \text{ to } l$$

where $a_j(k) \in \{0, 1\}$ is the contents of the jth memory element in the analyser in the kth cycle of its operation; $y(k) \in \{0, 1\}$ is the value of the binary digit applied to the analyser input in the kth cycle; $\alpha_i \in \{0, 1\}$ are coefficients defined by the form of $\varphi(x)$.

From equation (11.12) it follows that the contents of the first analyser's memory element in the $(k + 1)$th cycle is

$$a_1(k + 1) = y(k + 1) \oplus \alpha_1 y(k) \oplus \alpha_1 \sum_{i=1}^{m} \alpha_i a_i(k - 1) \oplus \sum_{i=2}^{m} \alpha_i a_{i-1}(k - 1)$$

and in the $(k + 2)$th cycle

$$a_1(k + 2) = y(k + 2) \oplus \alpha_1 y(k + 1) \oplus \alpha_1 \alpha_1 y(k) \oplus \alpha_1 \alpha_1 \sum_{i=1}^{n} \alpha_i a_i(k - 1)$$

$$\oplus \alpha_1 \sum_{i=2}^{m} \alpha_i a_{i-1}(k - 1) \oplus \alpha_2 y(k) \oplus \alpha_2 \sum_{i=1}^{m} \alpha_i a_i(k - 1) \oplus \sum_{i=3}^{m} \alpha_i a_{i-2}(k - 1).$$

In the general case, for the $(k + n - 1)$th cycle, we may write

$$a_1(k + n - 1) = \sum_{i=1}^{n} \beta_i(n) y(k - 1 + i) \oplus \sum_{i=1}^{m} \delta_i(n) a_i(k - 1) \qquad (11.13)$$

where $\delta_i(n) \in \{0, 1\}$, $i = 1$ to m, are coefficients that permit the formation of an n-cycle-shifted replica of an M-sequence defined by polynomial $\alpha(x)$. The value of $\delta_i(n)$ is obtained from the set of equations presented in [103]:

$$\delta_i(n) = \alpha_{i+n-1} \oplus \sum_{d=1}^{n-1} \alpha_d \delta_i(n - d) \qquad i = 1 \text{ to } m - n + 1, \ n = 1 \text{ to } m$$

$$\delta_i(n) = \sum_{d=1}^{n-1} \alpha_d \delta_i(n - d) \qquad i = m - n + 2 \text{ to } m, \ n = 1 \text{ to } m \qquad (11.14)$$

$$\delta_i(n) = \sum_{d=1}^{m} \alpha_d \delta_i(n - d) \qquad i = 1 \text{ to } m, \ n > m.$$

The values of $\delta_i(m)$, $i = 1$ to m, may also be obtained by fast formal procedures [47].
Coefficients $\beta_i(n) \in \{0, 1\}$ are given by [147]

$$\beta_i(n) = 0 \qquad i > n$$
$$\beta_i(n) = 1 \qquad i = n \qquad (11.15)$$
$$\beta_i(n) = \sum_{d=1}^{n-1} \alpha_d \beta_i(n - d) \qquad n > i, \ n - i \leqslant m$$
$$\beta_i(n) = \sum_{d=1}^{m} \alpha_d \beta_i(n - d) \qquad n - i > m.$$

From equation (11.13) for $a_i(k + i - 1)$ we can obtain its value by using n symbols $y(k), y(k + 1), \ldots, y(k + n - 1)$ of sequence $\{y(k)\}$ and m initial values $a_i(k - 1)$, $i = 1$ to m. At the same time, the above equation is used for constructing the functional diagram of a signature analyser that handles n symbols of $\{y(k)\}$ in each cycle. The analyser will then have n inputs such that it can be used to test a digital circuit with n inputs, and n of its output sequences are reduced to one of the form

$$y_1(k)y_2(k) \cdots y_n(k)y_1(k + 1)y_2(k + 1) \cdots y_n(k + 1)y_1(k + 2) \cdots y_n(k + 2) \ldots \quad (11.16)$$

where $y_v(k) \in \{0, 1\}$, $v = 1$ to n, is the value of the bit on the vth output of the circuit under test in the kth cycle of its operation. Then in accordance with equations (11.12) and (11.13) the behaviour of the analyser that handles the sequence of (11.16) may be described by the system

$$a_i(0) = 0 \qquad i = 1 \text{ to } m$$

$$a_1(k + n - 1) = \sum_{d=1}^{n} \beta_d(n) y_d(k) \oplus \sum_{i=1}^{m} \delta_i(n) a_i(k - 1)$$

$$
\begin{aligned}
a_j(k + n - 1) &= a_1(l + n - j) & j < n \\
a_j(k + n - 1) &= a_{j-n+1}(k - 1) & j \geqslant n
\end{aligned}
\qquad (11.17)
$$

$$k = 1 \text{ to } l \qquad j = 2 \text{ to } m.$$

Based on equation (5.27), a parallel analyser can be constructed to perform the same manipulations on the sequence (11.16) as the serial signature analyser for n cycles. The algorithm for constructing a parallel signature analyser consists of the following steps.

(i) For a specified probability P_n of failing to detect an error, the number m of memory elements in the signature analyser is defined by

$$P_n \geqslant 1/2^m. \qquad (11.18)$$

(ii) The generating polynomial $\varphi(x)$ is selected [87] and coefficients $\alpha_i \in \{0, 1\}$, $i = 1$ to m, are evaluated for $m = \deg \varphi(x)$.

(iii) The number of analyser inputs n determined by the number of circuit outputs is specified.

(iv) By using the system (11.14), constant coefficients $\delta_i(n + 1 - j) \in \{0, 1\}$, $i = 1$ to m, $j = 1$ to c, are computed for $c = \min(n, m)$.

(v) Coefficients $\beta_i(n + 1 - j) \in \{0, 1\}$, $i = 1$ to n, $j = 1$ to c, are found from the set of equations

$$\beta_n(n) = 1$$

$$\beta_{n-j}(n) = \sum_{d=1}^{\gamma} \alpha_d \beta_{n-j+d}(n) \qquad j = 1 \text{ to } n - 1, \gamma = \begin{cases} j, & j \leqslant m \\ m, & j > m \end{cases} \qquad (11.19)$$

$$\beta_{n-j-i+1}(n - j + 1) = \beta_{n-i}(n) \qquad i = 0 \text{ to } n - j, j = 1 \text{ to } c$$

obtained on the basis of equation (11.15).

(vi) The functional scheme of a parallel signature analyser is designed in accordance with the set of logic equations

$$a_j(k) = \sum_{d=1}^{n-j+1} \beta_d(n + 1 - j) y_d(k) \oplus \sum_{i=1}^{m} \delta_i(n + 1 - j) a_i(k - 1) \qquad j = 1 \text{ to } c$$

$$a_j(k) = a_{j-n}(k - 1) \qquad j = c + 1 \text{ to } m. \qquad (11.20)$$

Thus the procedure of parallel signature analyser design is considered to be complete.

Now we consider as an example the design of a PSA for testing digital circuits with the number of outputs not exceeding five and a probability $P_n = 0.000\,015$.

Table 11.2.

$n-j+1$	1	2	3	4	5	6	7	8	9	10	11	12	13	14	15	16
							i									
1	0	0	0	0	0	0	1	0	1	0	0	1	0	0	0	1
2	0	0	0	0	0	1	0	1	0	0	1	0	0	0	1	0
3	0	0	0	0	1	0	1	0	0	1	0	0	0	1	0	0
4	0	0	0	1	0	1	0	0	1	0	0	0	1	0	0	0
5	0	0	1	0	1	0	0	1	0	0	0	1	0	0	0	0

(i) Using equation (11.18) we find $m \geqslant 16$. To decrease hardware costs and shorten the signature length, let $m = 16$.

(ii) For $\deg \varphi(x) = 16$, we choose a generating polynomial $\varphi(x) = 1 \oplus x^7 \oplus x^8 \oplus x^{12} \oplus x^{16}$ whose coefficients $\alpha_i \in \{0, 1\}$, $i = 1$ to 16, when evaluated, are $\alpha_7 = \alpha_9 = \alpha_{12} = \alpha_{16} = 1$, $\alpha_1 = \alpha_2 = \alpha_3 = \alpha_4 = \alpha_5 = \alpha_6 = \alpha_8 = \alpha_{10} = \alpha_{11} = \alpha_{13} = \alpha_{14} = \alpha_{15} = 0$.

(iii) The number of analyser lines n is determined by the number of outputs in digital circuits under test, i.e. $n = 5$.

(iv) $c = \min(5, 16) = 5$. The estimates of coefficients $\delta_i(n - j + 1)$ obtained by using (11.14) are summarized in Table 11.2.

(v) By using equation (11.19), we obtain the estimates of $\beta_d(5)$ for $n = 5$

$$\beta_5(5) = 1$$
$$\beta_4(5) = \alpha_1 \beta_5(5) = 0$$
$$\beta_3(5) = \alpha_1 \beta_4(5) \oplus \alpha_2 \beta_5(5) = 0$$
$$\beta_2(5) = \alpha_1 \beta_3(5) \oplus \alpha_2 \beta_4(5) \oplus \alpha_3 \beta_5(5) = 0$$
$$\beta_1(5) = \alpha_1 \beta_2(5) \oplus \alpha_2 \beta_3(5) \oplus \alpha_3 \beta_4(5) \oplus \alpha_4 \beta_5(5) = 0.$$

The estimates of $\beta_d(n + 1 - j)$ for $d = 1$ to 5, $j = 1$ to 5, are summarized in Table 11.3.

(vi) The functional scheme of a five-input signature analyser is designed based on the set of equations

$$a_j(k) = y_{5+1-j}(k) \oplus a_{2+j}(k - 1) \oplus a_{4+j}(k - 1) \oplus a_{7+j}(k - 1)$$
$$\oplus a_{11+j}(k - 1) \qquad j = 1 \text{ to } 5$$
$$a_j(k) = a_{j-5}(k - 1) \qquad j = 6 \text{ to } 16 \qquad (11.21)$$

Table 11.3.

j	1	2	3	4	5
			i		
1	0	0	0	0	1
2	0	0	0	1	0
3	0	0	1	0	0
4	0	1	0	0	0
5	1	0	0	0	0

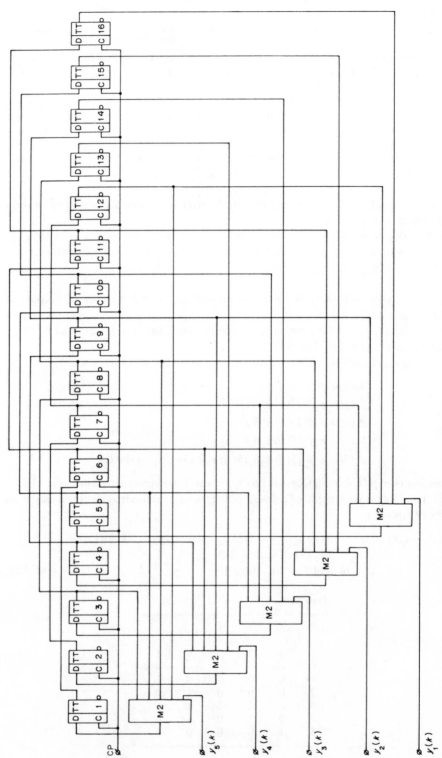

Fig. 11.4. Functional scheme of a five-input signature analyser for $\varphi(x) = 1 \oplus x^7 \oplus x^9 \oplus x^{12} \oplus x^{16}$

PARALLEL SIGNATURE ANALYSIS VALIDITY 181

Table 11.4.

			v		
k	1	2	3	4	5
1	1	0	1	0	1
2	0	1	0	1	1
3	1	1	1	0	1

Table 11.5.

										j						
k	1	2	3	4	5	6	7	8	9	10	11	12	13	14	15	16
0	0	0	0	0	0	0	0	0	0	0	0	0	0	0	0	0
1	1	0	1	0	1	0	0	0	0	0	0	0	0	0	0	0
2	1	1	1	1	0	1	0	1	0	1	0	0	0	0	0	0
3	1	0	0	1	1	1	1	1	1	0	1	0	1	0	1	0

derived from (11.20) by substituting the values of $\delta_i(5+1-j)$, $i=1$ to 5, and $\beta_d(5+1-j)$, $d=1$ to 5, $j=1$ to 5.

The set of equations (11.21) defines the connections between D flip-flops used as memory elements in the analyser of Fig. 11.4.

Consider the operation of the above parallel signature analyser when testing a five-input digital circuit that generates the sequences $\{y_v(k)\}$, $v=1$ to 5, $k=1$ to 3, that have been presented in Table 11.4 within the first three cycles.

Table 11.5 shows the states of the analyser's memory elements (Fig. 11.4) versus $y_v(k)$.

A comparison of the operation of the five-input signature analyser of Fig. 11.4 and that of a serial signature analyser described by $\varphi(x)=1\oplus x^7\oplus x^9\oplus x^{12}\oplus x^{16}$ shows that the states of the five-input analyser's memory elements repeat those of the serial signature analyser in five cycles. Then the five output sequences for a serial signature analyser are reduced to one subject to (11.16)

$$y_1(1)y_2(1)y_3(1)y_4(1)y_5(1)y_1(2)y_2(2)y_3(2)y_4(2)y_5(2)y_1(3)y_2(3)y_3(3)y_4(3)y_5(3).$$

The equivalent behaviour of parallel and serial analysers allows one to use the theory of serial analysers for estimating the efficiency of digital circuit analysis of parallel analysers.

11.3. ESTIMATION OF PARALLEL SIGNATURE ANALYSIS VALIDITY

Considering that an n-line signature analyser and its associated single-line analyser behave equivalently relative to the result of compressing n input sequences $\{y_v(k)\}$, $v=1$ to n, $k=1$ to l, we can estimate the validity of parallel signature analysis by using the results obtained while estimating the effectiveness of serial signature analysis. In fact, for a primitive polynomial $\varphi(x)$ the probability P_n of failing to detect an error in sequences $\{y_v(k)\}, v = 1$ to n, by an n-line signature analyser will be defined by

$$P_n = \frac{2^{nl-m}-1}{2^{nl}-1} \simeq \frac{1}{2^m} \tag{11.22}$$

for $nl = 2^m - 1$, where $m = \deg \varphi(x)$. This follows from the identity of the result obtained by the n-line signature analyser to the sequence compression result (11.16) obtained by the serial analyser. The expression (11.22) is true for any ratio of n and l whose product is $2^m - 1$. For $nl < 2^m - 1$, the value $1/2^m$ will be an average probability P_n of missing an error in sequences $\{y_v(k)\}$ being analysed. The given integral characteristic of the effectiveness of parallel signature analysis, like the characteristic of (8.4), is a rather crude estimation, which may be true only for general assumptions. A more complete characteristic for parallel signature analysis is the probability distribution P_n^μ of failing to detect a μ-multiple error in sequences $\{y_v(k)\}$ being analysed. Hence these probabilities, like those for serial signature analysis, may be estimated by equation (8.11), which holds true for the total sequence (11.16). An attempt to use equations (8.11) and (11.22) for estimating the effectiveness of analysing a single sequence $\{y_1(k)\}$ by a parallel analyser structure (see Sec. 11.2) may lead to

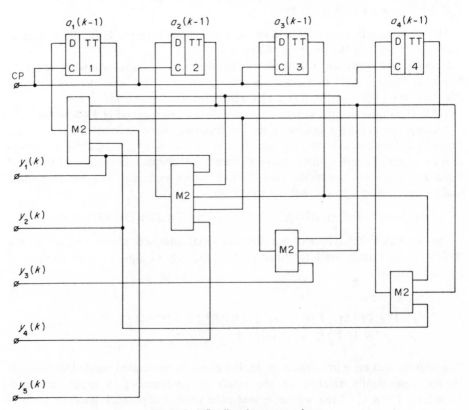

Fig. 11.5. A five-line signature analyser

contradictory conclusions. In the long run, this depends on specific parameters of the parallel signature analyser. In fact, probability P_n of failing to detect an error in the sequence $\{y_1(k)\}$, $k = 1$ to l, analysed by a serial analyser is determined by (8.4). The same sequence tested by parallel signature analyser together with sequences $\{y_v(k)\}$, $v = 2$ to n, will be determined by (11.22). Then for given equations (8.4) and (11.22) the condition

$$\frac{2^{nl-m}-1}{2^{nl}-1} < \frac{2^{l-m}-1}{2^l-1} \tag{11.23}$$

is satisfied, thereby implying that a parallel analyser scheme used for analysing the sequence $\{y_1(k)\}$ is preferable for achieving higher efficiency of the analysis. However, the above statement as well as relation (8.11) used as an estimate of probability P_n^μ of failing to detect an error dependent on its multiplicity μ may only be used for $nl \leqslant 2^m - 1$. An attempt to use the relation (8.11) for evaluating P_n^μ, on testing the sequence $\{y_1(k)\}$, $k = 1$ to l, with $l = 2^m - 1$, does not necessarily produce valid results in all applications of n-line analysis. To support this statement, consider as an example the five-line signature analyser based on polynomial $\varphi(x) = 1 \oplus x^3 \oplus x^4$ (Fig. 11.5). The set of logic equations obtained by the technique for designing such devices (see Sec. 11.2) and describing the connections of the analyser in question has the following form:

$$a_1(k) = y_1(k) \oplus y_2(k) \oplus y_5(k) \oplus a_2(k-1) \oplus a_4(k-1)$$
$$a_2(k) = y_1(k) \oplus y_4(k) \oplus a_1(k-1) \oplus a_3(k-1) \oplus a_4(k-1)$$
$$a_3(k) = y_3(k) \oplus a_1(k-1) \oplus a_2(k-1) \tag{11.24}$$
$$a_4(k) = y_2(k) \oplus a_2(k-1) \oplus a_3(k-1).$$

Let the above analyser be used for analysing the sequence $\{y_1(k)\}$ $k = 1$ to 15; $\{y_1(k)\} = 100110010100000$. For definiteness, assume that $\{y_v(k)\} = 000000000000000$ with $v = 2$ to 5. Sequences $\{y_v(k)\}$, $v = 1$ to 5, compressed by the analyser of Fig. 11.5 will result in the value of expected signature $S(y)$ that equals the contents $a_1(15)a_2(15)a_3(15)a_4(15)$ of its memory elements defined here by code 0001. When an error with multiplicity $\mu = 2$ that affects symbols $y_1(1)$ and $y_1(4)$ occurs, the actual sequence $\{y_1^*(k)\}$ will appear as $\{y_1^*(k)\} = 000010010100000$ and the value of signature $S^*(y)$ obtained by the five-line analyser will match its expected value $S(y) = 001$, thereby proving that the specified error has escaped detection, which contradicts the expression (8.11), according to which all double errors are detectable for $l \leqslant 2^m - 1$ where l is the length of sequence. The probability distribution P_n^μ is nonuniform, thereby suggesting that the quality of analysing the sequence $\{y_1(k)\}$ by the n-line analyser is lower than that of the classical single-line (serial) analyser. Equivalence of the analysis results obtained by two comparative structures of signature analysers on the sequence $\{y_1(k)\}$ will be reached only under condition $(n, l) = 1$, i.e. when n and l are relatively prime by the decimation property of the M-sequence [47]. We shall prove the statement for the case of $n = 2^d$, where d is a positive integer.

Theorem 11.1. Any set of errors in the sequence $\{y_1(k)\}$, $k = 1$ to $2^m - 1$, that are undetectable by a serial signature analyser implemented by the primitive polynomial $\varphi(x)$, whose deg $\varphi(x)$ is m, corresponds to the set of errors undetectable by the

$n = 2^d$ – line analyser described by polynomial $\varphi(x)$, provided sequences $\{y_q(k)\}, q = 2$ to n, and any positive integer d are errors-free.

Proof. The result of compressing n sequences $\{y_v(k)\}, v = 1$ to n, by the n-line signature analyser based on primitive polynomial $\varphi(x)$ corresponds to the result obtained by using the serial signature analyser for compressing the total sequence (11.16), which follows from the design rules for n-line signature analysers (see Sec. 11.2). Then the sequence of (11.16) can be represented as the binary polynomial $\kappa(x)$ which appears as

$$\kappa(x) = \bigoplus_{v=1}^{n} \bigoplus_{k=1}^{l} y_v(k)x^{(k-1)n+v-1}.$$

The given polynomial $\kappa(x)$, which corresponds to the reference sequences $\{y_v(k)\}$, $v = 1$ to n, may be represented as a modulo-2 sum of polynomial $\kappa^*(x)$, which describes the actual sequences $\{y_v^*(k)\}, v = 1$ to n, and error polynomial $e(x)$. Then the error polynomial $e(x)$ may be written like polynomial $\kappa(x)$ as

$$e(x) = \bigoplus_{v=1}^{n} \bigoplus_{k=1}^{l} e_v(k)x^{(k-1)n+v-1} \tag{11.25}$$

where $\{e_v(k)\}, v = 1$ to n, $k = 1$ to l, describe the errors that have occurred in sequences $\{y_i(k)\}$. By simplifying the expression (11.25) we obtain

$$e(x) = \bigoplus_{v=1}^{n} x^{v-1} \bigoplus_{k=1}^{l} e_v(k)x^{(k-1)n}$$

$$= \bigoplus_{v=1}^{n} x^{v-1} \bigoplus_{k=1}^{l} e_v^n(k)x^{(k-1)n}$$

$$= \bigoplus_{v=1}^{n} x^{v-1} \bigoplus_{k=1}^{l} [e_v(k)x^{k-1}]^n.$$

Considering that the sequences $\{y_d(x)\}, q = 2$ to n, are error-free, the expression for $e(x)$ will take the form

$$e(x) = \bigoplus_{k=1}^{l} [e_1(k)x^{k-1}]^n. \tag{11.26}$$

By using the Schönemann property [148], which holds for $n = 2^d$, where d is a positive integer, rearrange the expression (11.26) as

$$e(x) = \bigoplus_{k=1}^{l} [e_1(k)x^{k-1}]^n = \left(\bigoplus_{k=1}^{l} e_1(k)x^{k-1} \right)^n = e_1^n(x). \tag{11.27}$$

Here $e_1(x)$ is the polynomial that describes the errors that occur in $\{y_1(k)\}, k = 1$ to l, where $l = 2^m - 1$ and $m = \deg \varphi(x)$.

For the given polynomial the equality (8.2)

$$e_1(x) = g_1(x)\varphi^{-1}(x) \oplus S_1(x)$$

where $\varphi^{-1}(x)$ is the reciprocal of $\varphi(x)$ and $S_1(x)$ is the signature represented as a polynomial whose $\deg S_1(x) \leqslant m - 1$, holds true. A similar relation

$$e(x) = g(x)\varphi^{-1}(x) \oplus S(x) \tag{11.28}$$

also holds true for polynomial $e(x)$.

By using the relation (11.27), rearrange the expression (11.28) for polynomial $e(x)$ as

$$\begin{aligned}
e(x) = e_1^n(x) &= [g_1(x)\varphi^{-1}(x) \oplus S_1(x)]^n \\
&= g_1^n(x)[\varphi^{-1}(x)]^n \oplus S_1^n(x) \\
&= \{g_1^n(x)[\varphi^{-1}(x)]^{n-1} \oplus g_2(x)\}\varphi^{-1}(x) \oplus S(x)
\end{aligned}$$

where $S_1^n(x) = g_2(x)\varphi^{-1}(x) \oplus S(x)$. Hence the value of signature $S(x)$ generated by the n-line signature analyser for the sequence $\{y_1(k)\}$ is determined by the value of signature $S_1(x)$ for the same sequence and calculated by

$$S(x) = S_1^n(x) \bmod \varphi^{-1}(x). \tag{11.29}$$

Here $S_1(x)$ results from compressing $\{y_1(x)\}$ by a serial signature analyser described by a primitive polynomial $\varphi(x)$, which corresponds to the polynomial used for constructing the $n = 2^d$-line analyser. Then the set of errors in the sequence under test $\{y_1(k)\}$, which are undetectable by the serial signature analyser on analysing the value $S_1(x)$, corresponds to the set of errors undetectable by the n-line analyser, provided each non-zero value of $S_1(x)$ is associated with that of $S(x)$. In either case the actual value of signature obtained differs from the reference one, thereby proving that the error is undetectable by both signature analyser structures.

Next we shall demonstrate that the statement is true. For at purpose, we shall use the decimation property of M-sequences according to which decimation of an M-sequence by $n = 2^d$, where d is a positive integer, leads to a shifted replica of the initial M-sequence. Hence m symbols of signature $S_1(x)$ used for generation of polynomial $S_1^n(x)$ are independent symbols in the M-sequence described by $\varphi(x)$. Then each unit symbol of polynomial $S_1^n(x)$ will be associated with m coefficients that have been used for generation of the shifted M-sequence replica whose modulo-2 sum represents the value of $S(x)$ depending on the form of $S_1^n(x)$. Assuming that the unit symbols of polynomial $S_1^n(x)$ are independent, the modulo-2 sum of any set of coefficients that generate the values of these symbols of polynomial $S_1^n(x)$ from m initial symbols will be other than zero, i.e. $S(x) \neq 0$. Then zero value of $S(x)$ is only possible for linear dependence of the specified set of symbols in the polynomial $S_1^n(x)$, which is impossible for $n = 2^d$ by the decimation property of M-sequence and contradicts the assumptions. Thus any non-zero signature $S_1(x)$ is associated with a non-zero signature $S(x)$. Hence any error detected by a serial signature analyser may be detected by an n-line analyser and conversely any undetectable error that initiates a non-zero of $S_1(x)$ is undetectable by the n-line signature analyser, since for the zero polynomial $S_1(x)$, $S(x)$ will also be zero by (11.29). As a consequence, we may conclude that the set of errors in the sequence $\{y_1(k)\}$, $k = 1$ to $2^m - 1$, which are undetectable by the serial signature analyser, is associated with the set of errors that are undetectable by the $n = 2^d$-line analyser, as we wished to prove. ∎

Thus the efficiency of parallel signature analysis may be evaluated either by the integral value P_n, which is determined by (11.22), or by the distribution probability P_n^μ of failing to detect a μ-multiple error in the tested sequences $\{y_v(k)\}$, $v = 1$ to n, $k = 1$ to l, where $ln \leqslant 2^m - 1$ and $m = \deg \varphi(x)$. Hence the value of n determined as $n = 2^d$ is preferable. The analysis of sequence $\{y_1(k)\}$, $k = 1$ to l, for $l = 2^m - 1$, by such an analyser will be equivalent to the analysis by the appropriate serial signature analyser, thereby allowing the use of the parallel analyser in different modes. In so doing, its performance will not degrade in a serial mode as compared to the classical signature analyser structure.

In parallel with the high efficiency of analysis, the parallel signature analyser designed as described in Sec. 11.2 allows efficient implementation of the error sequence localization procedure. Let us consider some approaches to constructing such a diagnostic procedure.

11.4. APPLICATION OF PARALLEL ANALYSERS FOR FAULT DIAGNOSIS

By using parallel signature analysers we can significantly speed up the procedure of testing digital circuits, which is practically increased by n, where n is the number of inputs to the analyser applied. When the actual signature coincides with its reference value, the probability of the circuit under test being fault-free is high enough. Then the procedure of testing it stops. Otherwise, if the circuit is faulty the actual signature normally differs from the reference one, thereby suggesting that the circuit is faulty. At the same time the signature form obtained bears no additional information on the fault nature. The question of which of the n sequences under test initiating the actual signature contain faults remains open, i.e. it is necessary to isolate the fault to the sequence that carries the information on its existence. Consider the possible solutions to the task for the n-line analysers designed by the technique of Sec. 11.2. First, let us prove the following theorem.

Theorem 11.2. The total signature $S(x)$ obtained for sequences $\{y_v(k)\}$, $v = 1$ to n, $k = 1$ to l, by the n-line signature analyser equals the bit-by-bit modulo-2 sum of signatures $S_v(x)$, $v = 1$ to n, with each signature $S_j(x)$, $j \in \{1, 2, \ldots, n\}$, being generated for the sequence $\{y_j(k)\}$ if $\{y_q(k)\} = 00 \cdots 0$, $q \neq j \in \{1, 2, \ldots, n\}$.

Proof. According to equation (11.16) the equivalent input sequence tested by the n-line signature analyser can be described by the following binary polynomial:

$$\kappa(x) = \sum_{v=1}^{n} \sum_{k=1}^{l} y_v(k) x^{(k-1)k + v - 1} \tag{11.30}$$

which consists of the modulo-2 sum of two polynomials of the form

$$\kappa_j(x) = \sum_{k=1}^{l} y_j(k) x^{(k-1)n + j - 1} \qquad j \in \{1, 2, \ldots, n\}$$

describing input sequences $\{y_v(k)\}, v = 1$ to n. Each polynomial $\kappa_j(x)$ can be represented as a classical relation

$$\sum_{k=1}^{k} y_j(k)x^{(k-1)n+j-1} = g_j(x)\varphi^{-1}(x) \oplus S_j(x) \tag{11.31}$$

where polynomial $\varphi^{-1}(x)$ is the reciprocal of $\varphi(x)$ that is used for implementation of n-line signature analysis; $S_j(x)$ is a signature of sequence $\{y_j(x)\}$.

By modulo-2 summing the two right-hand and two left-hand parts of equation (11.31), we obtain that polynomial $\kappa(x)$ is defined as

$$\sum_{v=1}^{n} \sum_{k=1}^{l} y_v(k)x^{(k-1)n+v-1} = \sum_{v=1}^{n} g_v(x)\varphi^{-1}(x) \oplus \sum_{v=1}^{n} S_v(x)$$

for which equation (8.2)

$$\sum_{v=1}^{n} \sum_{k=1}^{l} y_v(k)x^{(k-1)n+v-1} = g(x)\varphi^{-1}(x) \oplus S(x)$$

also holds true.

A comparison of the two latter equations allows one to deduce that the total signature $S(x)$ obtained for the sequences $\{y_v(k)\}$, $v = 1$ to n, equals the bit-by-bit modulo-2 sum of signatures $S_v(x)$ for each input sequence

$$S(x) = \sum_{v=1}^{n} S_v(x) \tag{11.32}$$

as was to be proved. ∎

The above result expressed by (11.32) is true for a primitive polynomial $\varphi(x)$ and random values n and l. The corollary to the theorem is the possibility of finding the reference signature for a random set of input sequences. Thus the reference signature for the first, third and fifth sequences is defined by

$$S_{1,3,5}(x) = S_1(x) \oplus S_3(x) \oplus S_5(x).$$

Using the results of Theorem 11.2 we can formalize the procedure of testing a digital circuit on the basis of the n-input signature analyser. Then in the general case, the analyser's input sequences $\{y_v(k)\}$, $v = 1$ to n, $k = 1$ to l, may be the sequences produced at the input, intermediate and output nodes of the circuit for which the reference signatures $S_v(x)$, $v = 1$ to n, have been estimated in advance. Without loss of generality, let $n = 2^d$ and represent the testing procedure as the following algorithm.

The algorithm for testing a digital circuit with fault isolation to the first sequence comprising the errors caused by the fault is as follows.

(i) By analysing $n = 2^d$ actual sequences $\{y_v^*(k)\}$, $v = 1$ to n, estimate the signature $S^*(x)$ with an n-line analyser in accordance with

$$\sum_{v=1}^{n} \sum_{k=1}^{l} y_v^*(k)x^{(k-1)n+v-1} = g^*(x)\varphi^{-1}(x) \oplus S^*(x).$$

(ii) By

$$S(x) = \sum_{v=1}^{n} S_v(x)$$

obtain the reference estimate of $S(x)$.

(iii) The actual estimate of $S^*(x)$ is compared with the reference signature $S(x)$. When $S^*(x) = S(x)$, perform step (xi) and terminate the testing procedure. Otherwise, with $S^*(x) \neq S(x)$, proceed to the next step.

(iv) The whole set of input sequences is divided into two groups with the numbers of sequences $\{y_1(k)\}$, $\{y_2(k)\},\ldots,\{y_{n/2}(k)\}$ forming the set $A_1 = \{1, 2, \ldots, n/2\}$ and those of sequences $\{y_{n/2+1}(k)\}$, $\{y_{n/2+2}(k)\},\ldots,\{y_n(k)\}$ forming the set $A_2 = \{n/2 + 1, n/2 + 2, \ldots, n\}$; i is set to 1.

(v) The analysis of actual sequences whose numbers are specified by A_1 by an n-line analyser, provided that the sequences whose numbers are not specified by A_1 are zero, produces the estimate of actual signature $S^*(x)$.

(vi) On the basis of equation

$$S(x) = \sum_{j \in A_i} S_j(x)$$

we obtain $S(x)$.

(vii) Equation $S(x) = S^*(x)$ is checked for validity. When it holds true, the elements of A_1 are substituted by those of A_2.

(viii) The value of i is incremented by one and then compared with the value of d. With $i \leqslant d$ proceed to the next step, otherwise perform step (x).

(ix) Based on the current values of A_1 the new sets A_1 and A_2 are produced. As new elements of A_1 use the first half of its current elements, whereas its second half is assigned to set A_2. Having defined sets A_1 and A_2, proceed to step (v).

(x) A single element of A_1 is the number of the error sequence generated at a node of the circuit under test.

(xi) The procedure of digital circuit testing is considered complete.

As an example of the above algorithm, let us consider the case of testing the digital circuit of Fig. 11.6. For the purpose let us use an eight-line signature analyser that has been designed by the technique for parallel analysers (see Sec. 11.2). As a result, we obtain that the functional scheme of an eight-line analyser described by $\varphi(x) = 1 \oplus x^3 \oplus x^4$ corresponds to the set of logic equations

$$a_1(k) = a_1(k-1) \oplus a_2(k-1) \oplus a_3(k-1) \oplus a_4(k-1) \oplus y_2(k) \oplus y_4(k) \oplus y_5(k) \oplus y_8(k)$$
$$a_2(k) = a_2(k-1) \oplus a_3(k-1) \oplus a_4(k-1) \oplus y_1(k) \oplus y_3(k) \oplus y_4(k) \oplus y_7(k)$$
$$a_3(k) = a_1(k-1) \oplus a_3(k-1) \oplus y_2(k) \oplus y_3(k) \oplus y_6(k)$$
$$a_4(k) = a_2(k-1) \oplus a_4(k-1) \oplus y_1(k) \oplus y_2(k) \oplus y_5(k) \qquad k = 1 \text{ to } l.$$

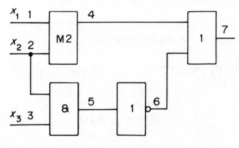

Fig. 11.6. A digital circuit

Table 11.6.

k	\multicolumn{8}{c}{v}
	1	2	3	4	5	6	7	8
1	0	0	0	0	0	1	0	0
2	0	0	1	0	0	1	0	0
3	0	1	0	1	0	1	1	0
4	0	1	1	1	1	0	0	0
5	1	0	0	1	0	1	1	0
6	1	0	1	1	0	1	1	0
7	1	1	0	0	0	1	0	0
8	1	1	1	0	1	0	0	0

Table 11.7.

$S_1(x)$	$S_2(x)$	$S_3(x)$	$S_4(x)$	$S_5(x)$	$S_6(x)$	$S_7(x)$	$S_8(x)$
0101	1100	0101	1110	1110	0000	0100	0000
5	F	5	P	P	0	4	0

Assuming that the initial state of the analyser's memory elements is always zero, i.e. $a_i(0) = 0$, $i = 1$ to 4, we estimate reference signatures $S_v(x)$, $v = 1$ to 8, for all nodes of the circuit under test. Sequences $\{y_v(k)\}$, $v = 1$ to 8, $k = 1$ to 8, generated at its nodes are summarized in Table 11.6. The values of $\{y_8(k)\}$, $k = 1$ to 8, are taken to be zero for the missing node 8 of the circuit of Fig. 11.6.

The estimates of all reference signatures in binary and hexadecimal are summarized in Table 11.7.

Using the reference signature $S_v(x)$ let us test a digital circuit with fault isolation to the first error sequence by the algorithm discussed above. Suppose that the circuit under test in Fig. 11.6 has a fault $f_6 \equiv 1$. As a result sequences $\{y_6(k)\}$ and $\{y_7(k)\}$ will be distorted to appear as

$$\{y_6^*(k)\} = \{y_7^*(k)\} = 1 \quad 1 \quad 1 \quad 1 \quad 1 \quad 1 \quad 1 \quad 1.$$

By successively performing the steps of the above algorithm, we obtain the following:

(i) The analysis of $8 = 2^3$ actual sequences

$$\{y_1^*(k)\} = 00001111 \qquad \{y_2^*(k)\} = 00110011$$
$$\{y_3^*(k)\} = 01010101 \qquad \{y_4^*(k)\} = 00111100$$
$$\{y_5^*(k)\} = 00010001 \qquad \{y_6^*(k)\} = 11111111$$
$$\{y_7^*(k)\} = 11111111 \qquad \{y_8^*(k)\} = 00000000$$

gives $S^*(x) = 1001 = 9$.

(ii) $S(x) = \sum_{v=1}^{8} S_v(x) = 0101 \oplus 1100 \oplus 0101 \oplus 1110 \oplus 1110 \oplus 0000 \oplus 0100 \oplus 0000 = 1000 = 8$.

(iii) Since $S^*(x) = 9 \neq S(x) = 8$, proceed to the next step.

(iv) Sets A_1 and A_2 are generated $A_1 = \{1, 2, 3, 4\}$ and $A_2 = \{5, 6, 7, 8\}$. The value of i is 1.

(v) The analysis of sequences

$$\{y_1^*(k)\} = 00001111 \qquad \{y_2^*(k)\} = 00110011$$
$$\{y_3^*(k)\} = 01010101 \qquad \{y_4^*(k)\} = 00111100$$

defined by A_1 results in a signature $S^*(x) = 0010 = 2$.

(vi) $S(x) = \sum_{j\in\{1,2,3,4\}} S_j(x) = 0101 \oplus 1100 \oplus 0101 \oplus 1110 = 0010 = 2$.

(vii) Since $S^*(x) = S(x) = 2$, the elements of A_1 are replaced with those of A_2. Therefore, $A_1 = \{5, 6, 7, 8\}$.

(viii) The value of i is set to 2. Since $i = 2 < d = 3$, proceed to the next step.

(ix) On the basis of $A_1 = \{5, 6, 7, 8\}$ sets $A_1 = \{5, 6\}$ and $A_2 = \{7, 8\}$ are generated.

(v') The analysis of sequences

$$\{y_5^*(k)\} = 00010001$$
$$\{y_6^*(k)\} = 11111111$$

results in $S^*(x) = 0010 = 2$.

(vi') $S(x) = \sum_{j\in\{5,6\}} S_j(x) = 1110 \oplus 000 = 1110 = P$.

(vii') $S^*(x) = 2 \neq S(x) = P$, therefore the elements of A_1 remain unchanged.

(viii') Since $i = d = 3$, proceed to the next step.

(ix') On the basis of $A_1 = \{5, 6\}$ the sets $A_1 = \{5\}$ and $A_2 = \{6\}$ are generated.

(v'') The analysis of $\{y_5^*(k)\} = 00010001$ results in $S_5^*(x) = 1110 = P$.

(vi'') $S(x) = \sum_{j\in\{5\}} S_j(x) = 1110 = P$.

(vii'') $S^*(x) = S(x) = P$, therefore the elements of A_1 are replaced with those of A_2 to result in $A_1 = 6$.

(viii'') With $i = 4$, proceed to step (x) after having analysed $i \leq d$.

(x) A unique element of A_1 is the number of the first error sequence. Thus all sequences whose numbers are less than that of A_1 element will correspond to reference values. For the example of interest $\{y_6^*(k)\} \neq \{y_6(k)\}$; hence $\{y_q^*(x)\} = \{y_q(x)\}$, $q = 1$ to 5.

(xi) The procedure of testing the discussed circuit is complete.

Thus from the analysis of the circuit of Fig. 11.6 we may deduce that its fault can distort the sequence at node 6 and probably at other nodes whose numbers belong to the set $\{7, 8\}$. The appropriate numbering of circuit nodes allows one to isolate the fault with high enough precision by the above algorithm. In the present case, the circuit fault is associated either with the NOT gate or its output node 6.

The algorithm discussed can be further extended to include the procedure of finding all sequences in error. This will make fault diagnosis more precise. The techniques for a serial signature analyser (see Sec. 9.4) can be used for implementing such an algorithm.

REFERENCES

[1] Lala, P. K. (1985) *Fault Tolerant and Fault Testable Hardware Design*, Prentice-Hall, Englewood cliffs, NJ.
[2] Bennetts, R. G. (1982) *Introduction to Digital Board Testing*, Crane-Russak, New York.
[3] Chegis, I. A. and Yablonsky, S. B. (1968) Logical testing techniques for electrical circuits. *Tr. Mat. Inst.*, p. 51.
[4] Karpovsky, M. (1983) Universal tests for detection of input/output stuck-at and bridging faults. *IEEE Trans. Comput.*, C-32, 1194–97.
[5] Parchomenko, P. P. (ed.) (1976) *Technical Diagnostic Fundamentals*, Energiya, Moskva.
[6] Chandramouli, R. (1983) On testing stuck-open faults. *Digest of Papers FTCS-13*, pp. 258–65.
[7] Timoc, C., Buehler, M. and Griswold, T. (1983) Logical models of physical failures. *Proc. IEEE. Test. Conf.*, pp. 546–53.
[8] Zakrevsly, A. D. (1984) *Logical Design of Cascaded Circuits*, Nauka, Moskva
[9] Abraham, J. A. and Fuchs, W. K. (1986) Fault and error models for VLSI. *Proc. IEEE*, 639–54.
[10] Galiay, J., Crouzet, Y. and Vergniault, M. (1980) Physical versus logical fault model in MOS LSI circuits: impact on their testability. *IEEE Trans. Comput.*, C-31, 527–531.
[11] Cha, C. W. (1978) A testing strategy for PLAS. *Proc. 15th Design Automation Conf.*, pp. 326–34.
[12] Somenzi, F. and Gai, S. (1986) Fault detection in programmable logic arrays. *Proc. IEEE*, 655–68.
[13] Komanytsky, D. (1983) Synthesis technique creates complete system for self-test. *Electronics*, 56, 5, 26–34.
[14] Bennetts, R. G. (1984) *Design of Testable Logic Circuits*, Addison-Wesley, Reading, MA.
[15] Bennetts, R. G. (1981) CAMELOT: a computer aided measure for logic testability. *Proc. IEEE*, 5, 177–89.
[16] Savir, J. (1981) Good controllability and observability do not guarantee good testability. *IEEE Trans. Comput.*, C-30, 55–61.
[17] Stephenson, J. E. Grason, J. (1976) A testability measure for register transfer level digital circuits. *Digest of Papers FTCS-6*, 101–17.
[18] Grason, J. (1979) TMEAS: a testability measurement program. *Proc. 16th Design Automation Conf.*, pp. 156–61.
[19] Goldstein, L. H. (1979) Controllability/observability analysis for digital circuits. *IEEE Trans. Circuits Systems*, CAS-26, 685–93.
[20] Ratiu, I. M. (1982) VICTOR: a fast VLSI testability analysis program. *Proc. IEEE Test. Conf.*, pp. 397–401.
[21] Armstrong, D. B. (1966) On finding a nearly minimal set of fault detection tests for combinatorial logic nets. *IEEE Trans. Electron. Comput.*, EC-15, pp. 66–73.
[22] Schneider, R. R. (1967) On the necessity to examine D-chains in diagnostic test generation. *IBM J. Res. Develop.*, 11, 114.
[23] Roth, J. P. (1966) Diagnosis of automata failure: a calculus and a method. *IBM J. Res. Develop.*, 10, 278–93.
[24] Chang, H. J., Manning, and E. Metzeg, G. (1970) *Fault Diagnosis of Digital Systems*, Wiley, New York.

[25] Breuer, M. A., Friedman, A. D. and Iosupovicz, A. (1981) A survey of the art of design automation. *IEEE Trans.* Comput., **C-30**, 58–75.

[26] Galynychev, V. N., Zoyagin, V. F. and Nemolochnov, O. F. (1981) A regular test sequence design method. Prefix test sequence. *Avtom. Telemech.*, **9**, 162–71.

[27] Sellers, F. F., Hsiao, M. Y. and Bearnson, C. L. (1975) Analyzing errors with the Boolean difference. *IEEE Trans. Comput.*, **C-29**, 62–71.

[28] Goldman, P. S. and Chipulis, V. P. (1976) *Technical Diagnostic of Digital Devices* Moskva Radio and Communication.

[29] Yamada, T., Saisho, M. and Kasuya, Y. (1979) Test generation method for highly sequential circuits. *Proc. COMPCON*, pp. 104–7.

[30] Cha, C. W. (1979) Multiple fault diagnosis incombinatorial networks. *Proc. 16th Design Automation Conf.*, pp. 149–55.

[31] Coryashko, A. P. (1984) Design of easily testable discrete devices: ideas, methods, implementation. *Avtom. Telemech.*, **7**, 5–33.

[32] Veitzman, I. N., Zuk, V. E. and Flerov, A. B. (1971) Probabilistic test generation system for logical circuits. *Voprosy Radioel., Ser. EVT*, **6**, 22–6.

[33] Agarwal, V. K. and Funy, A. S. F. 1981) Multiple fault testing of large circuits by single fault test sets. *IEEE Trans. Comput.*, **C-30**, 855–65.

[34] Reddy, S. M. (1972) Easily testable realization for logic functions. *IEEE Trans. Comput.*, **C-121**, 193–204.

[35] Dandapany, R. and Reddy, S. M. (1974) On the design of logic networks with redundancy and testability consideration. *IEEE Trans. Comput.*, **C-23**, 1139–49.

[36] Hayes, J. P. (1974) On modifying logic networks to improve their diagnosability. *IEEE Trans. Comput.*, **C-23**, 56–62.

[37] Saluia, K. K. and Reddy, S. M. (1974) On minimality testable logic networks. *IEEE Trans. Comput.*, **C-23**, 552–4.

[38] Yamada, A. (1977) Automatic system level test generation and fault location for large digital systems. *Proc. 14th Design Automation.*, pp. 347–52.

[39] Ando, H. (1980) Testing VLSI with random access scan. *Proc. COMPCON*, pp. 50–2.

[40] Eichelberger, E. B. and Williams, T. W. (1977) A logic design structure for LSI testability. *Proc. 14th Design Automation Conf.*, pp. 462–8.

[41] Williams, T. W. and Parker, K. P. (1982) Design for testability—a survey. *IEEE Trans. Comput.*, **C-31**, 2–15.

[42] McClusky, E. J. (1984) A survey of design for testability scan techniques. *VLSI Design*, **12**, 38–61.

[43] Godoy, H. C., Franclin, G. B. and Bottorff, P. S. (1977) Automatic checking of logic design structure for compliance with testability ground rules. *Proc. 14th Design Automation Conf.*, pp. 460–78.

[44] Boswell, F. R. (1972) Designing testability into complex logic boards. *Electronics Int.*, **17**, 113–21.

[45] Koenamann, B., Mucha, J. and Zwiehoff, G. (1980) Built-in test for complex integrated circuits. *IEEE J. Solid-State Circuits*, **SC-15**, 315–18.

[46] Mucha, J. (1981) Hardware techniques for testing VLSI circuits based on built-in test. *Proc. COMPCON*, pp. 366–9.

[47] Yarmolik, V. N. and Demidenko, S. N. (1988) *Generation and Application of Pseudorandom Sequences in Test and Check-out Systems*, Wiley, Chichester.

[48] Nading, H. J. (1977) Signature analysis—concepts, examples and guidelines. *Hewlett-Packard J.*, **24**(5), 43–9.

[49] Fasang, P. P. (1982) Circuit module implements practical self-testing. *Electronics Int.*, **6**, 164–7.

[50] Koenamann, B., Mucha, J. and Zwiehoff, G. 1979) Built-in logic block observation technique. *Proc. IEEE Test. Conf.*, pp. 37–41.

[51] Nesbet, S. B. and McCluskey, E. J. (1980) Structured design for testability to eliminate test pattern generation. *Digest of Papers FTCS-10*, pp. 158–63.

[52] McCluskey, E. J. (1982) Verification testing. *Proc. 19th Design Automation Conf.*, pp. 495–500.

[53] Fujiwara, H. and Kinoshita, K. (1978) Testing logic circuits with compressed data. *Digest of Papers FTCS-8*, pp. 108–13.

[54] Parker, P. K. (1976) Compact testing: testing with compressed data. *Digest of Papers FTCS-6*, pp. 93–8.

[55] Kazmina, S. K. (1982) Compact testing, *Avtom. Telemech.*, **2**, 137–89.

[56] Reddy, S. M., Saluja, K. K. and Karpovsky, M. A. (1985) Data compression technique for built-in self test. *Digest of Papers FTCS-15*, p. 299.

[57] Hayes, J. P. (1976) Transition count testing of combinational logic circuits. *IEEE Trans. Comput.*, **C-25**, 613–20.

[58] Savir, J. (1979) Syndrome-testable design of combinational circuits. *Digest of Papers FTCS-9*, pp. 137–40.

[59] Savir, J. (1980) Syndrome-testable design of combinational circuits. *IEEE Trans. Comput.*, **C-29**, 442–51.

[60] Barzilai, Z., Savir, J., Marcowsky, G. and Smith, M. G. (1981) The weighted syndrome sums approach to VLSI testing. *IEEE Trans. Comput.*, **C-30**, 996–1000.

[61] Parker, P. K. and McCluskey, E. J. (1975) Analysis of logic circuits with faults using input signal probabilities. *IEEE Trans. Comput.*, **C-24**, 574–8.

[62] Bastin, D. (1973) Probabilistic test generation techniques. *Digest of Papers FTCS-3*, p. 171.

[63] Bernstein, M. S. and Romankevich, A. M. (1974 Statistical methods of checking logic circuits. *Kibernetika*, **1**, 52–7.

[64] Bernstein, M. S. and Romankevich, A. M. (1975) On statistical checking of multi-pin circuits. *Avtom. Vych. Tekh.*, **1**, 14–20.

[65] Frohwerk, R. A. (1977) Signature analysis: a new digital field method. *Hewlett-Packard J.*, **28**, 9, 1–8.

[66] Reddy, S. M. (1977) A note on testing logic circuits by transition counting. *IEEE Trans. Comput.*, **3**, 313–14.

[67] Savir, J. (1981) Syndrome-testing of syndrome-untestable combinational circuits. *IEEE Trans. Comput.*, **C-30**, 606–8.

[68] Markowsky, G. (1981) Syndrome-testability can be achieved by circuit modification. *IEEE Trans. Comput.*, **C-30**, 604–6.

[69] Carter, W. C. (1982) The ubiquitous parity bit. *Digest of Papers FTCS-12*, pp. 289–96.

[70] Suskind, A. K. (1981) Testing by verifying Walsh coefficients. *Digest of Papers FTCS-11*, pp. 206–8.

[71] Teller-Giron, R. and David, R. (1974) Random fault-detection in logical networks. *Digest of Papers, Int. Symp. on Discrete Systems, Riga*, pp. 232–41.

[72] Yakovlev, V. V. and Fedorov, R. F. (1974) *Stochastic Computers*, Mashinostroyenie, Leningrad.

[73] Parker, K. P. and McCluskey, E. J. (1975) Probabilistic treatment of general combinatorial networks. *IEEE Trans. Comput.*, **C-24**, 668–70.

[74] Agrawal, P. and Agrawal, V. D. (1976) On Monte Carlo testing of logic tree networks. *IEEE Trans. Comput.*, **C-25**, 664–7.

[75] Agrawal, P. and Agrawal, V. D. (1975) Probabilistic analysis of random test generation method for irredundant combinational logic networks. *IEEE Trans. Comput.*, **C-24**, 691–5.

[76] Savir, J., Ditlow, G. S. and Bardell, P. H. (1984) Random pattern testability. *IEEE Trans. Comput.*, **C-33**, 79–90.

[77] Jain, S. K. and Agrawal, V. D. (1985) Statistical fault analysis. *IEEE Design Test. Comput.*, Febr., 38–44.

[78] Eichelberger, E. B. and Lindbloom, E. (1983) Random-pattern coverage enhancement and diagnosis for LSSD logic self-test. *IBM J. Res. Develop.*, **27**, 265–72.

[79] Williams, T. W. (1985) Test length in a self-testing environment. *IEEE Design Test. Comput.*, April, 59–63.

[80] Tamamoto, H. and Narita, Y. (1982) A method for determining optimal or quasi-optimal input probabilities in random test method. *Denski Tsushin Gakkai Ronbunshi*, **65**, 8, 1057–64.

[81] Virupakshia, A. R. and Reddy, V. C. (1983) A simple random test procedure for detection of single intermittent fault in combinational circuits. *IEEE Trans. Comput.*, **C-32**, 594–7.

[82] Agrawal, V. D. (1978) Whento use random testing. *IEEE Trans. Comput.*, **C-27**, 1054–5.

[83] Knuth, D. (1977) *The Art of Computer Programming*, vol. 2. *Seminumerical Algorithms*, Addison-Wesley, Reading, MA.

[84] Yarmolik, V. N., Leusenko, A. E. and Demidenko, S. N. (1981) Investigation of linear congruential sequence properties. *Autom. Explor. Invest., Mansk*, **2**, 118–23.

[85] Peterson, W. W. (1961) *Error Correcting Codes*, Wiley, New York.

[86] Lawis, T. G. and Payne, W. H. (1973) Generalized feedback shift register pseudorandom number algorithm. *J. ACM*, **20**, 456–68.

[87] Stahnke, W. (1973) Primitive binary polynomials. *Math. Comput.*, **27**, 277–80.

[88] Arvillias, A. C. and Maritsas, D. G. (1979) Toggle-registers generating in parallel k-th decimations of M-sequences $x^r + x^i + 1$: design tables. *IEEE Trans. Comput.*, **C-28**, 2, 89–101.

[89] Sarwate, D. V. and Pursley, M. V. (1980) Cross correlation properties of pseudorandom and related sequences. *Proc. IEEE*, **68**, 593–619.

[90] Willett, M. (1976) Characteristic M-sequences. *Math. Comput.*, **30**, 306–11.

[91] Alekseev, A. I., Sheremetev, A. G., Tuzov, G. I. and Glasov, B. I. (1969) *Theory and Application of Pseudorandom Signals*, Nauka, Moscow.

[92] MacWilliams, F. J. and Sloane, N. I. (1976) Pseudorandom sequences and arrays. *Proc. IEEE*, **64**, 1715–29.

[93] Badulin, S. S., Barnaulov, V. A., Berdyshev, V. A., *et al.* (1981) *Automated Design of Digital Devices*, Radio i Svyaz, Moscow.

[94] Yarmolik, V. N. (1983) A test signal sequence generator for probabilistic testing of digital devices. *Electron Modelling*, **5**, 49–54.

[95] Yarmolik, V. N. (1985) Hardware test pattern generators. *Izv. Vuzov. Priborostr.*, **27**, 35–40.

[96] Lidak, V. Ju., Podunajev, G. A., *et al.* (1983) Automated system for testing and diagnosis of logic units at the production stage. *Avtom. Vych. Tekh.*, **3**, 57–63.

[97] Borshchevich, V. I., Klistorin, I. F.and Filimonov, S. N. (1985) Simulation of the process for generation of pseudorandom tests of digital devices by means of stochastic grammars. *Elektron. Model.*, **2**, 53–7.

[98] Sidorenko, O. D., Rukkas, E. N., *et al.* (1982) Organization of processes of digital block check and diagnostics. *Upr. Sistemy Mashiny*, **4**, 18–22.

[99] Bernstein, M. S., Karachun, L. F. and Romankevich, A. M. (1978) Pseudorandom signal sequence generator. *Managem. Mech. Autom.*, **1**, 57–60.

[100] Yarmolik, V. N. (1982) Pseudorandom test signal sequence generator. *Managem. Mech. Autom.*, **3**, 57.

[101] Romankevich, A. M. (1978) On a method for constructing nonlinear generators of pseudorandom sequences. *Kibernetika*, **1**, 136–7.

[102] Romankevich, A. M. and Karachun, L. F. (1981) Towards the estimation of characteristics of test sequences in a probability system of technical diagnostics. *Upr. Sistemy Machiny*, **3**, 33–5.

[103] Yarmolik, V. N. (1983) Design of high-speed parallel pseudorandom number generators. *Izv. Vuzov. Priborostr.*, **26**, 48–52.

[104] Yarmolik, V. N. and Morozevich, A. N. (1977) A design method for multichannel high-speed pseudorandom number generators. *Radio Electron. Eng., Minsk*, 7, 66–9.

[105] Yarmolik, V. N. (1983) Designing generators of pseudorandom sequences of test signals. *Avtom. Telemekh.*, **6**, 155–72.

[106] Yarmolik, V. N. and Demidenko, S. N. (1984) Generation of M-sequences for automated test and verification systems. *Autom. Manuf. Design, Minsk*, **1**, 116–22.

[107] Chamitov, G. P. (1983) *Random Process Simulation*, Izd-vo Irkutsk. Un-ta, Irkutsk.

[108] Fedorov, R. F., Yakovlev, V. V. and Dobris, G. U. (1978) *Stochastic Data Converters*, Mashinostroyenie, Leningrad.

[109] Gladky, V. S. (1973) *Probabilistic Computation Models*, Nauka, Moscow.

[110] Yarmolik, V. N., Leusenko, A. E. and Morozevich, A. N. (1982) Random process generation by simple digital circuits. *Izv. Vuzov. Priborostr.*, **25**, 31–4.

[111] Tang, D. T. and Chen, C. L. (1984) Logic test pattern generation using linear codes. *IEEE Trans. Comput.*, **C-33**, 845–9.

[112] Tang, D. T. and Chen, C. L. (1984) Iterative exhaustive pattern generation for logic testing. *IBM J. Res. Develop.*, **28**, 212–19.

[113] Tang, D. T. and Chen, C. L. (1983) Logic test pattern generation using linear codes. *Digest of Papers FTCS-13*, pp. 190–4.

[114] Barzilai, Z., Coppersmith, D. and Rosenberg, A. L. (1983) Exhaustive generation of bit patterns with applications to VLSI self-testing. *IEEE Trans. Comput.*, **C-32** 190–4.

[115] Yarmolik, V. N. (1986) Integrated LSI self-test. *Mikroelektronika*, **15**, 1, 70–6.

[116] Hsiao, M. Y., Patel, A. M. and Pradhan, D. K. (1977) Store address generator with on-line fault-detection capability. *IEEE Trans. Comput.*, **C-26**, 1144–7.

[117] Wang, L. T. (1982) Autonomous linear feedback shift register with on-line fault-detection capability. *Digest of Papers FTCS-12*, pp. 311–14.

[118] Tang, D. T. and Woo, L. S. (1983) Exhaustive test pattern generation with constant weight vectors. *IEEE Trans. Comput.*, **C-32**, 1145–50.

[119] Eiki, H., Inagaki, K. and Yajima, S. (1980) Autonomous testing and its application to testable design of logic circuits. *Digest of Papers ETCS-10*, pp. 173–8.

[120] Litikov, I. P. (1983) Ring-wise testing of combinational circuits. *Avtom. Telemekh.*, 7, 145–53.

[121] Latypov, R. Kh. (1984) On reliability of ring-wise testing of linear sequential circuits. *Avtom. Vych. Tekh.*, **4**, 89–91.

[122] Smirnov, N. I., Struchkov, A. A. and Sudovtsev, V. A. (1979) Fault diagnosis in digital radioelectronic equipment based on LSI. *Zarub. Radioelektr.*, **1**, 53–60.

[123] Kir'yanov, K. G. (1980) On the theory of signature analysis. Communications equipment. *Radioizm. Tekh.*, **2**(27), 1–46.

[124] Ioffe, M. I. (1984) Estimating the capability of signature analysis to detect errors of specified multiplicity in a binary sequence. *Avtom. Telemekh.*, **12**, 110–18.

[125] Fujiwara, H. (1985) *Logic Testing and Design for Testability*, MIT Press, Cambridge, MA.

[126] Smith, I. F. (1980) Measures of the effectiveness of fault signature analysis. *IEEE Trans. Comput.*, **C-29**, 510–14.

[127] Novik, G. Kh. (1982) On the credibility of signature analysis. *Avtom. Telemekh.*, **5**, 157–9.

[128] Baran, E. D. (1982) Reliability of testing of binary sequences by the method of state counting. *Avtom. Vych. Tekh.*, **6**, 66–70.

[129] Yarmolik, V. N. (1985) On the validity of binary data sequence testing by signature analyzer. *Elektron. Model.*, **6**, 49–57.

[130] Latypov, R. Kh. (1985) Signature analysis comparison against a trivial compression technique at linear combinational circuit fault detection. *Avtom. Telemekh.*, **2**, 165–7.

[131] Katznelson, E. I. and Yarmolik, V. N. (1986) An approach to estimating the compact testing method efficiency. *Avtom. Projektir. Mikroprotses. Ustrojstv.*, *Minsk*, pp. 59–62.

[132] Yarmolik, V. N. (1984) Use of signature analysis for testing and diagnostics of complex digital integrated structure. *Avtom. Vych. Tekh.*, **4**, 73–80.

[133] Yarmolik, V. N. and Katznel'son, E. I. (1986) Validity of checksum test method and signature analysis for testing and diagnostics of discrete network structures. *Avtom. Vych. Tekh.*, **3**, 80–6.

[134] David, R. (1978) Feedback shift register-testing. *Digest of Papers FTCS-8*, pp. 103–7.

[135] David, R. (1980) Testing by feedback shift register. *IEEE Trans. Comput.*, **C-29**, 668–73.

[136] David, R. (1984) Signature analysis of multi-output circuits. *Digest of Papers FTCS-14*, pp. 366–71.

[137] Susskind, A. K. (1981) Testing of verifying Walsh coefficients. *IEEE Trans. Comput.*, **C-30**, 198–201.

[138] Muzio, J. C. and Miller, D. M. (1982) Spectral techniques for fault detection. *Digest of Papers FTCS-12*, pp. 297–302.

[139] Miller, D. M. and Muzio, J. C. (1983) Spectral fault signature for internally unate combinational networks. *IEEE Trans. Comput.*, **C-23**, 1058–64.

[140] Karpoosky, M. (1986) *Finite Orthogonal Series in the Design of Digital Devices*, Wiley, New York.

[141] Eris, E. and Miller, D. M. (1983) Syndrome-testable internally unate combinational networks. *Electron. Lett.*, **19**, 637–9.

[142] Saxena, N. R. and Robinson, J. P. (1985) Accumulator compression testing. *Digest of Papers FTCS-15*, pp. 300–5.

[143] Saluja, K. K. and Karpovsky, M. (1983) Testing computer hardware through data compression in space and time. *Proc. IEEE Test. Conf.*, pp. 83–8.

[144] MacWilliams, F. J. and Sloane, N. J. A. (1978) *The Theory of Error Correcting Codes*, North-Holland, Amsterdam.

[145] Sridhar, T., Ho, D. S., Powell, T. J. and Thatle, S. M. (1982) Analysis and simulation of parallel signature analyzers. *Proc. IEEE Test Conf.*, **C-31**, 656–61.

[146] Hassan, S. Z., Lu, D. J. and McCluskey, E. J. (1983) Parallel signature analyzers— detection capability and extensions. *Proc. COMPCON*, pp. 440–5.

[147] Yarmolik, V. N. (1985) Design of multi-channel signature analyzers *Avtom. Telemekh.*, **1**, 127–32.

[148] Faradzev, R. G. (1975) *Linear Sequential Machines*, Soviet Radio, Moscow.

INDEX